FIRST COURSE IN MATHEMATICAL LOGIC

A Blaisdell Book
in the Pure and Applied Sciences

ROBERT E. K. ROURKE
Saint Stephen's School, Rome

CONSULTING EDITOR

FIRST COURSE IN MATHEMATICAL LOGIC

PATRICK SUPPES, STANFORD UNIVERSITY

and

SHIRLEY HILL, UNIVERSITY OF MISSOURI

Blaisdell Publishing Company

A DIVISION OF GINN AND COMPANY

Waltham, Massachusetts · Toronto · London

PREFACE

In modern times logic has become a deep and broad subject. Only in recent years have systematic relations between logic and mathematics been established and a completely explicit theory of inference formulated which is adequate to deal with all the standard examples of deductive reasoning in mathematics and the empirical sciences. The concept of axioms and the derivation of theorems from axioms is at the heart of all modern mathematics. The purpose of this book is to introduce the student to modern mathematics at a level which is rigorous yet simple enough in presentation and context to permit relatively easy comprehension.

The importance of both the theory of proof and the methodology of deriving theorems from axioms in modern mathematics cannot be questioned. Yet development of the skills of deductive reasoning has been left largely to incidental learning in the school mathematics curriculum. The point of view represented in this book is that deliberate and well-planned teaching of mathematical logic when presented early in the career of a student will provide background for a deeper and more penetrating study of mathematics.

The scope of the present volume comprises the sentential theory of inference, inference with universal quantifiers, and applications of the theory of inference developed to the elementary theory of commutative groups, or, as it is put in the text, to the theory of addition. Because of the complexities introduced by existential quantifiers, their consideration has been delayed and will be given in the subsequent volume, *Second Course in Mathematical Logic*. It may be noted that the restriction to universal quantifiers occurring at the beginning of formulas is not as severe as it may seem. Most elementary mathematical theories which the student will encounter may be easily formulated within this framework. This restriction provides the student with an opportunity to learn how to make rigorous and nontrivial mathematical proofs without becoming entangled with the subtleties that surround existential quantification. Emphasis throughout the book has also been given to the pervasive and important problem of translating English sentences into logical or mathematical symbolism.

v

The present book is the fourth version of a set of notes developed in 1960–61. The second version of the text was used with eleven classes of selected students in 1961–62. The third version was used experimentally in 1962–63 with ten classes of selected school students and 200 college students in a project sponsored jointly by the Office of Education and the National Science Foundation. The writing of the book has also been supported by grants from the Carnegie Corporation of New York.

We have attempted to write the book in such a fashion that it may be used by students in a wide range of age and ability. Logic, fortunately, is one of the subjects not requiring extensive background or experience in order to be thoroughly mastered. For this reason a book of this particular type may be used by a wide variety of students. We believe that the book will prove useful for many a secondary school and elementary college mathematics class. The *Second Course in Mathematical Logic* is in preparation for those classes with time for a more ample exposure to the subject.

We are grateful to Mrs. Madeline Anderson for her patient and competent work in typing the manuscript. We are most indebted to Mr. Frederick Binford for many helpful suggestions and criticisms. He has also been responsible for preparing the detailed Teacher's Edition. Mr. Richard Friedberg gave a number of very useful comments and criticisms on the final draft of the manuscript.

PATRICK SUPPES
SHIRLEY HILL

Stanford University
Stanford, California
January, 1963

CONTENTS

FIRST COURSE IN MATHEMATICAL LOGIC

CHAPTER ONE
SYMBOLIZING SENTENCES

▶ 1.1 *Sentences*

In the study of logic our goal is to be precise and careful. The language of logic is an exact one. Yet we are going to go about building a vocabulary for this precise language by using our sometimes confusing everyday language. We need to draw up a set of rules that will be perfectly clear and definite and free from the vagueness we may find in our natural language. We can use English sentences to do this, just as we use English to explain the precise rules of a game to someone who has not played the game. Of course, logic is more than a game. It can help us learn a way of thinking that is exact and very useful at the same time.

To begin, let us look at English sentences. Each sentence has a logical form to which we shall give a name. At first, we shall be discussing and symbolizing two kinds of sentences in logic. The names given to these two types of sentences are *atomic* and *molecular*.

In this age of science, you have seen the word *atomic* used many times. As a matter of fact, the meaning of the word in the language of logic is similar to its original meaning in the physical sciences. In logic, *atomic* refers to the simplest (or most basic) kind of sentence. If we put one or more atomic sentences together with a connecting word, then we have a *molecular* sentence. An atomic sentence is one complete sentence with no connecting words. We use connecting words to make molecular sentences from atomic sentences.

For example, let us consider two atomic sentences,

Today is Saturday.
There will be no school.

Both sentences are atomic sentences. By using a connecting word, we can put them together and we will have a molecular sentence. For example, we can say

Today is Saturday and there will be no school.

1

This molecular sentence is made up of two atomic sentences and the connecting word 'and'. When we take a molecular sentence apart, we separate it into its smallest complete atomic sentences. In the example above, we can separate the molecular sentence into the two atomic sentences. The connecting word 'and' is not a part of either atomic sentence. It is added to the atomic sentences to make a molecular sentence.

▶ 1.2 *Sentential Connectives*

The connecting words, small as they may be, cannot be overlooked for they are very important. In fact, we shall learn some strict rules for the use of these key words. Much of what we shall be doing in our study of logic depends upon how carefully these connecting words are used. The connecting word in the sample sentence 'Today is Saturday and there will be no school' is the word 'and'. There are others, but before we take up each word separately we should learn the correct logical name for them. That name is *sentential connective*. This name should be very easy to remember because it actually tells us what job the word does. It connects sentences. It makes molecular sentences from atomic sentences.

The sentential connectives we shall use in this chapter are the words 'and', 'or', 'not', and 'if . . . , then . . . '. In the study of English you may learn other names for them, but in learning logic we shall call them all *sentential connectives*, or just *connectives*. Remember that when you add a sentential connective to either one or two atomic sentences, you then have formed a molecular sentence. Three of the above connectives, 'and', 'or', and 'if . . . , then . . . ', are used to connect two atomic sentences, but one of them is added to just one atomic sentence to make a molecular sentence. That connective is the word 'not'. We may say then that the connective 'not' controls *one* atomic sentence at a time and that the other connectives control *two* atomic sentences at a time. Remember that the connective 'not' is the only one that does not really connect *two* sentences. When added to just a single sentence, "not" forms a molecular sentence.

Let us look at some examples of molecular sentences that use the connectives we have named. The sentence,

The moon is not made of green cheese

is a molecular sentence that uses the connective 'not'. The connective, in this case, controls just *one* atomic sentence: 'The moon is made of green cheese'.

An example of a sentence using the connective 'or' is

Those clouds will be blown away or it will surely rain today.

The connective 'or' controls *two* atomic sentences. They are 'Those clouds will be blown away' and 'It will surely rain today'. The molecular sentence,

If this is October then Halloween is coming soon

illustrates the use of the connective 'if . . . , then . . . ', which also controls *two* atomic sentences. Can you tell what those two atomic sentences are? We have already seen an example of a sentence that uses the connective 'and'. Another is

The soil is very rich and there is enough rainfall.

What are the two atomic sentences contained in this molecular sentence?

The exercises below will give you a chance to check your ability to recognize atomic sentences, molecular sentences, and sentential connectives. Remember that every sentence having a connective is molecular.

E X E R C I S E 1

A. Write an A for each sentence that is an atomic sentence and an M for each sentence that is a molecular sentence. For each molecular sentence write the sentential connective used.

1. Lunch will be at exactly noon today.
2. The big black bear lumbered lazily down the road.
3. That music is very soft or the door is closed.
4. That big dog likes to chase cats.
5. He called for his pipe and he called for his bowl.
6. Bob is a good player or he is very lucky.
7. If Bob is a good player, then he will be on the school team.
8. California is west of Nevada and Nevada is west of Utah.
9. Many college students study logic during the first year.
10. Kittens do not usually wear mittens.

11. If kittens wear mittens, then cats may wear hats.
12. Jane can be found at Susan's house.
13. Sea lions do not grow long manes.
14. If Molly is singing, then she must be happy.
15. Sophomores do not follow freshmen in the registration schedule.
16. Jack's favorite subject is mathematics.
17. If those clouds are moving this way, then we will have rain.
18. If wishes were horses, then beggars would ride.
19. This sentence is atomic or it is molecular.
20. The sun was hot and the water looked very inviting.
21. If $x = 0$ then $x + y = 1$.
22. $x + y > 2$.
23. $x = 1$ or $y + z = 2$.
24. $y = 2$ and $z = 10$.

B.   Make four molecular sentences by using one or two of the atomic sentences listed below together with a sentential connective. For example, you could put the connective 'and' between two of them. You may use the same atomic sentence more than once. Use each of the four connectives *exactly once* so that each of your molecular sentences has a different connective.

1. The wind blows very hard.
2. Paul should be able to win easily.
3. The rain may cause them to call off the race.
4. We shall know what the plans are by tomorrow.
5. There will still be time to get there by seven.
6. Jean's friend is right.
7. We were wrong about the time for the meeting.

C.   Tell what the connectives are in each of the following sentences. For each molecular sentence, tell the number of atomic sentences you find in it. Remember that 'if . . . then . . .' is a single connective.

1. This is not my lucky day.
2. The winter is coming and the days grow shorter.
3. Many germs are not bacteria.
4. Amphibians are found in fresh water or they are found on land near moist places.
5. If there are fractures in great rock masses then earthquakes are likely to occur.
6. That number is greater than two or it is equal to two.

7. If it is a positive number then it is greater than zero.
8. This boy is my brother and I am his sister.
9. My score is high or I shall get a low grade.
10. If you hurry then you will be on time.
11. If $x>0$ then $y=2$.
12. If $x+y=2$ then $z>0$.
13. $x=0$ or $y=1$.
14. If $x=1$ or $z=2$ then $y>1$.
15. If $z>10$ then $x+z>10$ and $y+z>10$.
16. $x+y=y+x$.

D. First make up five atomic sentences and then make up five molecular sentences.

▶ 1.3 *The Form of Molecular Sentences*

The rules for the use of the sentential connectives are the same no matter what atomic sentences they connect or with what atomic sentences they are used. In one of the last exercises you found that it was possible to choose any one or two of a group of atomic sentences and combine them with a connective. The *form* of the molecular sentences that you made up depended upon the connective you chose, not upon what was in the atomic sentence or sentences. In other words, if you replace the atomic sentences in a molecular sentence with any other atomic sentences the form of the molecular sentence will remain the same. The way the 'if . . . , then . . .' connective is written shows this clearly. The three dots after 'if' and the three dots after 'then' stand in place of sentences. To form a molecular sentence using that connective you can simply put atomic sentences, *any* atomic sentences, in place of the dots.

It is easy to see the *form* of a molecular sentence if we do not write out the atomic sentences but just show where they belong. We can show the form of a molecular sentence using the 'and' connective as

$$\text{_____ and _____}$$

or as

$$(\qquad) \text{ and } (\qquad)$$

Any sentences can be put in the spaces and the form is the same. Suppose we chose the two atomic sentences 'It is red' and 'It is blue'.

Filling in the spaces above, we have the molecular sentence, 'It is red and it is blue'. We might have chosen two other atomic sentences and formed, for example, the sentence, 'I am tall and he is small'. The form remains the same. It is a molecular sentence using the connective 'and'. Another way of emphasizing the form is to leave the parentheses in the English sentence, as in the following sentences:

(It is red) and (it is blue).
(It is raining) and (Peter is wet).

We have said that we can fill in the spaces with any sentences. We are not limited to just atomic sentences. We can also use molecular sentences and the *form* is the same. For example, we can fill the first space with the molecular sentence, 'John is not here' and the second space with the molecular sentence 'Herb is not here'. The sentence will then be

John is not here and Herb is not here.

Again, the form is the same. The connective 'and' connects two sentences but this time they are molecular sentences.

We could also have used one molecular sentence and one atomic sentence, as in

John is not here and Joe is here.

The important point is that whatever sentences we use to fill the spaces shown, the form of the sentence is that of a molecular sentence using the connective 'and'.

This is true of the other connectives too. We might show the form of other types of molecular sentences as follows:

(   ) or (   ).
If (   ) then (   ).

We can fill the spaces with any sentences, atomic or molecular. The following are examples, some of which have parentheses included for emphasis.

Mary is here or Helen is home.
(John is in town) or (Mary is not at home).
If $2+3=x$ then $x=5$.
If $(y+1=4)$ then $(y=3)$.
If (Bill is not dishonest) then (John is honest).

Sometimes, in English sentences, we use one word for a particular connective, sometimes we use two or more. For instance, we can use the single word 'or' as a connective as in

> It is very heavy or it is hollow.

We might also write the same sentence adding the word 'either' as a part of the connective.

> Either it is very heavy or it is hollow.

The words 'either' and 'or' are both part of the connective. In English sentences we sometimes use 'either-or' and sometimes just 'or'. When we refer to the connective 'or' we know that this may also include the word 'either', if we choose to use it. The form for the connective 'or', therefore, may be

> Either (   ) or (   ).

The following examples are of this form.

> Either John is here or it is not raining.
> Either (Mary is not here) or (Susan is not here).
> Either $x+y=6$ and $y=2$, or $x=0$.
> Either $(x+y=7$ and $y \neq 2)$ or $(x>0)$.

In some cases, we may wish to include the word 'both' in addition to 'and' as a way of using the 'and' connective. For example, we can say

> Both it is raining and the sun is shining.

The words 'both' and 'and' are parts of the connective. We usually just use 'and' but you may find the word 'both' included occasionally. We shall refer to the connective as 'and', but the form may be seen as

> Both (   ) and (   ).

For example,

> Both $(x>0)$ and $(y \neq 0)$.
> Both $x \neq y$ and $y \neq z$.

In most cases in which the connective 'if . . . , then . . .' is used, both words are included. Sometimes, however, you may find that the word 'then' is eliminated. An example might be

> If it is Dan, he is late.

Sentences of this kind are formed by the 'if . . . , then . . .' connective and are of the form

If (   ), (   ).

Examples of this form are

If $x+y=2$ and $y=0$, $x=2$.

If $(x+y=7$ and $x=6)$, $(y=1)$.

If Mary loves John, John loves Mary.

The word 'not' is found most often within the atomic sentence in English. For this reason it is easy to overlook. But a sentence such as

Logic is not difficult

is a molecular sentence because it contains 'not'. It is possible to write the connective using the phrase 'it is not the case that'. The sentence then would read

It is not the case that logic is difficult.

Thus it is possible to show the form of the molecular sentence using the connective 'not' as

It is not the case that (   )

or we may shorten it to

not (   ).

Examples of this form are

It is not the case that $(x=0)$.

It is not the case that $(x+y>2)$.

Not $(x=2+1)$.

Not $(7>x+y)$.

The use of 'Not (   )' is not ordinary English usage, of course, but it is often convenient, as we shall see later in mathematical contexts.

In mathematical sentences using the equality symbol $=$ we often indicate negation by a slant line through the equality symbol: $\neq$. Thus, '$x\neq 1$' is read 'x is not equal to 1'.

In neither of the sentences '$x\neq 1$' and 'John is not here', can we use parentheses to show the form of the molecular sentence because the connective 'not' occurs inside the atomic sentence.

E X E R C I S E 2

A. Use parentheses to show the form of the following molecular sentences.

1. John is here and Mary has left.
2. If $x+1=10$ then $x=9$.
3. Either Mary is not here or Jane is gone.
4. If either $x=1$ or $y=2$ then $z=3$.
5. If $x\neq1$ and $x+y=2$ then $y=2$.
6. If either Smith is at home or Jones is in court then Scott is innocent.
7. $y=0$ and $x=0$.
8. Either $y=0$ and $x\neq0$ or $z=2$.
9. It is not the case that $6=7$.
10. It is not the case that if $x+0=10$ then $x=5$.

B. Write English sentences in the following forms. Drop the parentheses when you write the English sentences.

1. Either (   ) or (   ).
2. (   ) or (   ).
3. Both (   ) and (   ).
4. (   ) and (   ).
5. Not (   ).
6. If (   ) then (   ).
7. If (   ), (   ).
8. If not (   ) then not (   ).
9. It is not the case that (   ).

▶ 1.4 *Symbolizing Sentences*

We often think of atomic sentences as short sentences. But even some of the atomic sentences in our everyday language are long and for this reason they become clumsy and awkward to handle. In logic we take care of this problem by using *symbols* in place of, or to stand for, entire sentences.

The symbols we use in logic to stand for sentences are capital letters such as 'P', 'Q', 'R', 'S', 'A', and 'B'. For example, let

P = 'The snow is deep',
Q = 'The weather is cold'.

Now consider the sentence, 'The snow is deep and the weather is cold'. First, we show the logical form of the sentence by using parentheses.

(The snow is deep) and (the weather is cold.)

Then we use 'P' and 'Q' to symbolize the sentence as

(P) and (Q).

Suppose we wish to symbolize a molecular sentence that uses the connective 'or'. Consider the sentence 'You may choose soup or you may choose a salad'. We symbolize it in the following way:
Let

R = 'You may choose soup',
S = 'You may choose a salad'.

The sentence will then be symbolized,

(R) or (S).

In symbolizing a sentence that includes the connective 'not', the word 'not' is put in front of the symbol that stands for the atomic sentence even though in the English sentence you will usually find the word 'not' inside the atomic sentence that it controls. The connective, however, is not a part of the atomic sentence and therefore the word 'not' must be separated from the atomic sentence. For example, we would symbolize the sentence 'Ducks are not four-legged animals' in this way:
Let

Q = 'Ducks are four-legged animals'.

The molecular sentence will then be

Not (Q).

The letter symbol stands only for the atomic sentence and does not include the connective.

You will find later that if we use symbols for the atomic sentences it is far easier to deal with the molecular sentences, which can become very long and complicated.

The following exercises will give you practice in symbolizing sentences.

EXERCISE 3

A. Symbolize the following molecular sentences by replacing the atomic sentences with capital letters.

1. I need to put on my glasses or this light is poor.

Let

$$G = \text{'I need to put on my glasses'},$$
$$L = \text{'This light is poor'}.$$

Then the sentence is symbolized as

$$(\textbf{G}) \text{ or } (\textbf{L}).$$

2. Ducklings do not grow into swans.
3. He walked three steps to the right and then he went two steps forward.
4. These problems are not easy for me.
5. If the buzzer sounds, then it is time for class to begin.
6. If chemistry class has already begun then I am late.
7. One side of the moon is not seen from the earth.
8. Either Dick will go to the dance or he will go to the show.
9. Roses are red and violets are blue.
10. If Brazil is in South America then it is in the Southern Hemisphere.

B. Translate the following sentences into English sentences that have the same form. (Use the same connective and replace the letters with atomic sentences.) Tell the atomic sentence for which each letter symbol stands.

1. If (P) then (Q)	6. Not (P)
2. (R) or (S)	7. (R) and (T)
3. (P) and (Q)	8. (S) or (Q)
4. Not (E)	9. Not (T)
5. If (S) then (B)	10. If (R) then (S)

C. Each of the following sentences is molecular. First tell what the connective or connectives are in each sentence. Then write out every atomic sentence you find in each of the molecular sentences.

1. John is second and Tim is fourth.
2. Either Jack is the winner or Jim is the winner.
3. Eddie is not the winner.
4. If Jim is the winner then he gets the medal.
5. If Jim is not the winner then he must have placed second.
6. The Alps are young mountains and the Appalachians are old mountains.
7. Spiders are not insects.
8. If spiders are insects then they must have six legs.
9. If material is heated then it expands.
10. Most planets are either too hot for living things like ourselves or they are too cold for living things like ourselves.

D.   Symbolize the following mathematical sentences by replacing the atomic sentences with capital letters. Remember that $\neq$ is the negation of $=$.

1. If $x=y$ then $x=2$.
2. If $x \neq 2$ then $y>1$.
3. If either $x \neq 2$ or $x \neq 3$ then $x=1$.
4. If $x+y=3$ then $y+x=3$.
5. If $x-y=2$ then $y-x \neq 2$.
6. $x+y=2$ and $y=1$.
7. $x+y+z=2$ or $x+y=10$.
8. If $x \neq y$ and $y \neq z$ then $x>z$.
9. If $x+y>z$ and $z=1$ then $x+y>1$.
10. If $x \neq y$, then $x \neq 1$ and $x \neq 2$.

▶ 1.5 *The Sentential Connectives and Their Symbols*

Now that we can symbolize atomic sentences, working with molecular sentences will be much easier. But we can also use symbols for the sentential connectives themselves. We shall take up each of the connectives separately and learn the symbol for it. We shall also learn a name for the molecular sentence that is formed by the use of each connective. These connecting words are so important that we shall discuss them separately in the following sections, reviewing some of the things we have already discussed.

And.   When we use the word 'and' to join two sentences we call this the *conjunction* of two sentences. An example of a conjunction is this sentence,

Your eyes are brown and your brother's eyes are brown too.

Let **P** be the atomic sentence 'Your eyes are brown' and let **Q** be the atomic sentence 'Your brother's eyes are brown too'. Then we may symbolize the molecular sentence, which is a conjunction, by

(**P**) and (**Q**).

A conjunction is a type of molecular sentence. The molecular sentence is the conjunction of atomic sentence **P** and atomic sentence **Q**. It is also helpful to have a symbol for 'and'. We use the symbol that occurs on most typewriters,

&.

Using this symbol, we may write the conjunction of two sentences **P** and **Q** as

(**P**) & (**Q**).

Remember that the symbol & stands for the entire connective whether it is 'and' or 'both . . . and . . .' in English. The symbol & is called the ampersand.

EXERCISE 4

A.   Symbolize completely the following sentences, using the proper logical symbol for the connectives. Show the atomic sentence for which each capital letter stands.

1. Bob lives on our street and Janet lives on the next block.
2. Joe's old records are good and the newer ones are even better.
3. He put in his thumb and he pulled out a plum.
4. The sun went behind a cloud and at once it seemed cooler.
5. The jet climbed out of sight and it left a faint, white trail behind.
6. Janet is thirteen and Pat is fifteen.
7. George is tall and Andy is short.
8. Starfish are echinoderms and sand dollars are also echinoderms.

9. Today's date is the thirtieth and tomorrow will be the first.

10. The game has begun and we shall be late.

B.   Finish symbolizing the following sentences by replacing the connective with its proper logical symbol.

1. (P) and (Q)
2. Both (A) and (B)
3. (H) and (K)
4. Both (T) and (G)
5. (S) and (Q)

C.   Translate the sentences below into English sentences that have the same form. This means that English sentences will be substituted for letter symbols and the English connective substituted for the logical symbol.

1. (P) & (Q)
2. (R) & (S)
3. (T) & (C)
4. (B) & (H)
5. (Q) & (P)

D.   In the following mathematical sentences, symbolize only the connective 'and'.

1. $x=0$ and $y=4$.
2. $x\neq0$ and $x+y=2$.
3. $x-x=0$ and $x+0=x$.
4. $x+y=y+x$ and
$x+(y+z)=(x+y)+z$.

Or.   When we use the word 'or' to connect two sentences, we call this the *disjunction* of the two sentences. For example,

This is Room Four or this is a physics classroom

is the *disjunction* of two sentences. A disjunction is a molecular sentence formed by the connective 'or'. The molecular sentence above may seem a little clumsy to you. Probably this is because in our everyday language we often include the word 'either' when we use the word 'or'. For example, our molecular sentence might read

Either this is Room Four or this is a physics classroom.

In either case, the two atomic sentences are still the same; first, the sentence 'This is Room Four' and second, the sentence 'This is a physics classroom'. In other words, do not make the mistake of including the word 'either' as a part of the first atomic sentence. It is a part of the connective.

The symbol we will use for disjunction is the wedge sign,

$$\lor.$$

In the preceding example if we let **F** be the sentence 'This is Room Four' and **R** be the sentence 'This is a physics classroom', then the whole disjunction is symbolized

$$(\mathbf{F}) \lor (\mathbf{R}).$$

We can read this sentence by saying (**F**) *or* (**R**). Sometimes we may say *Either* (**F**) *or* (**R**). Remember that the symbol $\lor$ stands for the entire connective, whether the words in the English sentence are 'or' or 'either . . . , or . . .'.

EXERCISE 5

A.   Symbolize completely the following sentences, using the proper symbol for the connectives. Tell the atomic sentence for which each capital letter stands.

1. The area of triangle *ABC* is equal to the area of triangle *DEF* or the area of triangle *ABC* is less than the area of triangle *DEF*.
2. He will enter the high jump event or he will run the half mile.
3. Either she will act in the play or she will help with the costumes.
4. Either the boat sprang a leak or it was swamped by the waves.
5. We will have to get there earlier or someone else will get the job.
6. Either the needle is worn out or the record is a bad one.
7. Either Jones will be re-elected or he will be appointed to a new position.
8. We can specify the vector by giving two components, or our concern is three dimensions.
9. Lungfishes can take oxygen from air or they can take oxygen from water.
10. Either an anemone is an animal or it is a plant.

B.   Finish symbolizing the following sentences by putting in the proper symbol for the connective.

1. (**P**) or (**Q**)
2. Either (**P**) or (**Q**)
3. Either (**R**) or (**S**)
4. (**T**) or (**E**)
5. Either (**P**) or (**N**)

C.   Translate the following sentences into English sentences of the same form.

1. (P) ∨ (Q)
2. (R) ∨ (S)
3. (G) ∨ (H)

4. (R) ∨ (Q)
5. (A) ∨ (E)

D.   Symbolize the following mathematical sentences using & and ∨ but retaining the mathematical symbols.

1. Either $x=0$ or $x>0$.
2. $x\neq0$ and $y\neq0$.
3. Either $x>1$ or $x+y=0$.
4. Either $y=x$ or $y\neq x$.
5. $y+x>y+x+z$ or $z=0$.
6. $y+z=z+y$ and $0+x=x$.

E.   Symbolize the following mathematical sentences using & and ∨ but retaining the mathematical symbols and the parentheses.

1. Either $(x+y=0$ and $z>0)$ or $z=0$.
2. $x=0$ and $(y+z>x$ or $z=0)$.
3. Either $x\neq0$ or $(x=0$ and $y>0)$.
4. Either $(x=y$ and $z=w)$ or $(x<y$ and $z=0)$.

Not.   When the connective 'not' is added to a sentence, we shall call the result the *negation* of the sentence. Thus, a negation is a molecular sentence using the connective 'not'. The connective 'not' is like the other connectives in that it makes molecular sentences out of atomic sentences. But it is unlike the other connectives because it is used with only a *single* sentence. The word 'not' is usually found within a sentence in everyday language. In logic, however, we should learn to think of the connective apart from the sentence it controls. This is necessary in order to translate a negation into logical symbols.

An example of a negation is the sentence

Presidential elections do not often end in ties.

Although this resembles a simple atomic sentence because it contains only one atomic sentence, it is not. It is the negation of the atomic sentence,

Presidential elections do often end in ties.

In logic the addition of the connective 'not' to an atomic sentence results in a molecular sentence. Because in ordinary language we are most likely to meet a negation with the word 'not' *inside* the atomic sentence, it is easy to make the mistake of forgetting to place the 'not' in front of the capital letter we choose to symbolize the atomic sentence. The correct way of symbolizing the sentence, 'Presidential elections do not often end in ties', would be this

Let

$$P = \text{'Presidential elections do often end in ties'}$$

Then the sentence is written

$$\text{Not } (P).$$

In order to symbolize the sentence completely, we use a symbol for the negation,

$$\neg.$$

The symbolized sample sentence is now

$$\neg(P).$$

It is sometimes easier to think of the sentence in English as beginning with the phrase 'It is not the case that'. You may always translate the symbol $\neg$ as 'it is not the case that'. For example, to translate the sentence $\neg(P)$ back into the English sentence about Presidential elections, we might say, 'It is not the case that Presidential elections do often end in ties'.

Connectives may be used with one or more molecular sentences as well as atomic sentences. For example, in the form '*If* (   ) *then* (   )', we can fill the empty spaces with either atomic or molecular sentences. Negations are often combined with other sentences to form a longer molecular sentence. For example,

If a number is greater than 0, then it is not a negative number

is a molecular sentence of the 'if . . . then . . . ' form in which the connective joins an atomic sentence and a negation. The form, 'Either (   ) or (   )' might include negations as in the following disjunction:

Either the game has not started or there is not a big crowd.

Here we have the disjunction of two molecular sentences, both negations. We symbolize this sentence in the same way that we do other molecular sentences. First, its logical form can be shown more clearly by putting parentheses in the English sentence,

Either (the game has not started) or (there is not a big crowd).

Choosing a capital letter for each atomic sentence, we then show its negation by putting the symbol ¬ in front of the letter. Next, we connect the two molecular sentences by our major connective, which in this case is the 'or' connective. Completely symbolized, the sentence might look like this:

$$(\neg S) \lor (\neg C).$$

EXERCISE 6

A. Symbolize completely the following sentences, using the proper symbol for the connectives. Show the atomic sentence for which each capital letter stands.

1. In the Southern Hemisphere, July is not a summer month.
2. Neon tubes are not incandescent.
3. It is not the case that all income is taxed at the same rate.
4. Mars is not closer to the sun than the earth is.
5. Texas is not the largest state in the United States.
6. It is not the case that all liquids boil at the same temperature.
7. John Quincy Adams was not the second President of the United States.
8. Not all germs are bacteria.
9. It is not the case that a sea lily is a flower.
10. Louise is not a tall person.

B. Symbolize completely the following sentences by using the proper symbol for each connective.

1. It is not the case that (R)   4. It is not the case that (T)
2. Not (Q)   5. Not (J)
3. Not (H)

C. In the following sentences, more than one connective is used. Symbolize the sentences completely by replacing connectives by the proper symbol.

1. (P) and not (Q)   4. Either not (P) or not (Q)
2. Not (R) and not (M)   5. (T) and not (R)
3. (S) or not (B)

D.   First, write out each connective in the sentences below. Then symbolize the entire sentence, letting **P**='Jack is early' and **Q**='Tim is late' in all five sentences.

1. Either Jack is early or Tim is late.
2. Either Jack is not early or Tim is late.
3. Tim is late and Jack is not early.
4. Tim is not late and Jack is not early.
5. Jack is not early and Tim is late.

E.   Identify each of the following molecular sentences by writing the word that denotes its form (for example, 'negation', 'conjunction', 'disjunction').

1. ¬(**Q**)
2. (**P**) & (**Q**)
3. ¬(**R**)
4. (**R**) ∨ (**S**)
5. (**R**) & (**S**)
6. ¬(**T**)
7. (**P**) ∨ (**Q**)
8. (**R**) & (**T**)
9. ¬(**S**)
10. (**T**) ∨ (**Q**)

F.   Examine the following sentences and write out every connective you find in each.

1. It is not noon and lunch is not ready.
2. If we are not there then we lose our vote.
3. If two numbers are not equal then one is greater than the other.
4. Mary has gone or she is not at her desk.
5. If it is black then it will not reflect the light.
6. $x>0$ or $x=0$.
7. If $x+y=z$ then $y+x=z$.
8. If $x+y=0$ and $x>0$, then $y<0$.
9. If $x+y=0$ and $x=0$, then $y=0$.
10. Either $x=0$ or $x\neq0$.

If . . . then. . . .   When we put two sentences together by using the words 'if . . . then . . .' to join them, we call the resulting molecular sentence a *conditional sentence.* Earlier we said that the way the connective 'if . . . then . . .' is written indicates the form of the conditional sentence. Any sentence can be put in place of the dots. The word 'if' comes before the first sentence and the word 'then' comes before the second sentence.

An example of a conditional sentence is

If it rains today, then the picnic will be postponed.

The first atomic sentence is 'It rains today' and the second atomic sentence is 'The picnic will be postponed'. In order to symbolize this conditional sentence completely we use the following symbol for the connective:

$$\rightarrow.$$

We are now able to symbolize our sentence in the following way. First, we select capital letters for the atomic sentences. Let

P = 'It rains today'
Q = 'The picnic will be postponed'.

Then we write in the symbol for the connective.

$$(\textbf{P}) \rightarrow (\textbf{Q}).$$

There are some logical terms for the parts of a conditional sentence which you should learn. The sentence that comes between the word 'if' and the word 'then' is the *antecedent*. The sentence that follows the word 'then' is the *consequent*. We shall be using these terms often when we work with conditional sentences.

EXERCISE 7

A. Symbolize the following sentences, using the proper symbol for the connectives. Show the atomic sentence for which each capital letter stands.

1. If it is cold enough, then the lake will be frozen over.
2. If the lights are on, then the Smiths are at home.
3. If two pulses pass through each other, then they continue on in their original shapes.
4. If you miss the bus, then you will have to walk.
5. If you drive northward, then you will reach Canada tomorrow.
6. If it is an acid, then it contains the element hydrogen.
7. If two plus three equals five, then three plus two equals five.
8. If x is equal to two, then x plus one is equal to three.
9. If this is the seventh, then Friday is the ninth.
10. If his production increases, then Jones will keep the price stable.

B. Examine the following conditional sentences and write out the antecedent in each.

1. If Janet is youngest then Molly is oldest.
2. If Molly is oldest then Karen is younger.
3. If Janet is youngest then Mike is older.
4. If Mike is older then he is seventeen.
5. If Mike is seventeen then Karen is seventeen.

C. Examine the following conditional sentences and write out the consequent in each.

1. If Ted is second then Bob is third.
2. If Bob is third then he is before John.
3. If John is fourth then Carl is fifth.
4. If Ted is second then he is after Mark.
5. If Ted is after Mark then Mark is first.

D. Symbolize completely the following sentences by replacing the connectives by logical symbols.

1. If (P) then (R)
2. If (S) then (T)
3. If (Q) then (P)
4. If (P) then not (S)
5. If not (S) then not (T)

E. Identify the conditional sentences among the sentences below by putting the letter C after each sentence of that form.

1. $(P) \lor \neg(Q)$
2. $(P) \rightarrow \neg(Q)$
3. $(R) \rightarrow (S)$
4. $(T) \mathbin{\&} (S)$
5. $(T) \mathbin{\&} \neg(S)$
6. $(T) \rightarrow (S)$
7. $(R) \lor (P)$
8. $(R) \rightarrow (P)$
9. $(Q) \rightarrow (S)$
10. $(R) \mathbin{\&} (T)$

▶ 1.6 *Grouping and Parentheses*

We have learned that it is common to find sentences having more than one connective. Connectives may connect or be used with molecular sentences as well as atomic sentences. In all such cases, one of the connectives is the *major* connective. We may also call it the *dominant* connective because it dominates the sentence.

Recall that the form of one type of molecular sentence can be shown as

$$(\quad) \, \& \, (\quad).$$

This is a conjunction and the spaces may be filled with atomic or molecular sentences. Now, if molecular sentences are used, then they in turn contain connectives. But the & remains the dominant or major connective. Suppose we have the conjunction of two negations as in the sentence,

Tony is not in high school and Ann is not in high school.

If we let **T** be the sentence 'Tony is in high school' and **A** be the sentence 'Ann is in high school', the sentences that would then fit the form given above are ¬**T** and ¬**A**.

$$(\neg \mathbf{T}) \, \& \, (\neg \mathbf{A}).$$

Consider a conjunction whose left member is itself a disjunction and whose right member is an atomic sentence. This means that the 'and' connective connects a molecular sentence formed by using 'or' with an atomic sentence.

Both $x=1$ or $x=2$, and $y=3$.

If we let **P**=`$x=1$', **Q**=`$x=2$', and **R**=`$y=3$'; then the disjunction is (**P**) ∨ (**Q**) and the atomic sentence is **R**. When these sentences are filled into the proper spaces for a conjunction, the result is

$$(\,(\mathbf{P}) \lor (\mathbf{Q})\,) \, \& \, (\mathbf{R}).$$

With so many parentheses close together this symbolic sentence is difficult to read. So, we adopt the convention that *a sentence containing no &, ∨, or → need not be enclosed in parentheses.* This means that we do not need parentheses around either the '**P**' or the '**Q**'. Accordingly, the previous symbolic sentence may be written as

$$(\mathbf{P} \lor \mathbf{Q}) \, \& \, (\mathbf{R}).$$

But since '**R**' also contains no &, ∨, or →, the sentence reduces to

$$(\mathbf{P} \lor \mathbf{Q}) \, \& \, \mathbf{R}.$$

We can quickly see that this is a conjunction. The connective 'and' joins two sentences. One is the atomic sentence **R**; the other is a molecular sentence, the disjunction, **P** ∨ **Q**.

Parentheses are the punctuation symbols of logic. They show the way in which a sentence is grouped and therefore tell us which connective is dominant. With parentheses around P ∨ Q, we know that the parts are tied together as a single sentence. The molecular sentence can be joined to something else by a connective just as an atomic sentence can.

Notice that in the English sentence which we symbolized above, the same job is done by the comma. But suppose the sentence had read

Either $x = 1$, or $x = 2$ and $y = 3$.

Here the comma shows us that the major connective should be the 'or'. Since the form of the disjunction is

$$(\quad) \vee (\quad)$$

we fill the spaces with an atomic sentence and a conjunction:

(P) ∨ (Q & R).

Notice that without parentheses the two symbolized sentences look just alike. Because of the reasons given previously, we do not need the parentheses around the atomic sentence; therefore the sentence in final form is

P ∨ (Q & R).

When we symbolize English sentences, we need some way of picking out the major connective in the sentence. Although the parentheses always make the major connectives very clear in logic, they are not always as clear in English sentences because there are different methods of indicating dominance. One method, as we have seen, is the use of commas.

The clearest method of showing the dominance of a connective in English is to use the two-word form of the connective to surround the molecular sentence.

Both (   ) and (   ).
Either (   ) or (   ).
If (   ) then (   ).

For example, consider this English sentence:

(1)   Either he is wrong and I am right or I shall be
         surprised.

Inserting parentheses, we obtain

> Either (he is wrong and I am right) or (I shall be
> surprised).

The parentheses make it clear that the words 'either' and 'or' surround the conjunction 'he is wrong and I am right' which is just one part of the whole disjunction. Thus sentence (1) can be symbolized

(2) $(W \ \& \ R) \lor S.$

On the other hand, if the parentheses are placed so that the & is outside, then it will dominate and the whole sentence then becomes a conjunction,

(3) $W \ \& \ (R \lor S).$

The English for this would be

(4)   He is wrong and either I am right or I shall be
           surprised.

Notice the different positions of the word 'either' in the two English sentences (1) and (4). When 'either' comes first the disjunction dominates as in (1) and (2); when 'either' comes later the disjunction does not dominate as in (3) and (4).

It is possible to put in the 'both' to match the 'and'. By inserting parentheses to make the form clear, sentence (1) would then be

(5)   Either (both he is wrong and I am right) or (I shall
           be surprised)

but it would still be a disjunction symbolized by formula (2). And sentence (4) with parentheses would be

(6)   Both (he is wrong) and (either I am right or I shall be
           surprised)

which would still be a conjunction symbolized by formula (3).

Putting in every 'both' and every 'either' may make awkward English, and so we do not always include them, but it makes the logical form very clear. When these words are used, the first word of the sentence tells what logical type of sentence it is: 'both' shows it is a conjunction with 'both . . . and . . .' dominant, 'either' shows it is a disjunction with 'either . . . or . . .' dominant, and 'if' shows it is a conditional with 'if . . . then . . .' dominant. To make better sounding

English 'both', 'either', and 'then' can often be left out and the sentence still may have clear meaning. Unfortunately this is sometimes done when they are needed and then it is impossible to decide what the sentence means. It is then ambiguous like

(7)   He is wrong and I am right or I shall be surprised.

We cannot be sure whether (7) is a disjunction or a conjunction.

EXERCISE 8

Copy sentences (1), (4), (5), (6), and (7) and try in each sentence to put the parentheses in different places. It cannot be done in (1), (4), (5), and (6), but can in (7). This shows (1), (4), (5), and (6) are clear with only one meaning but that (7) is ambiguous because it has more than one possible meaning.

When we have mathematical sentences to translate into logical symbols, the same methods can be used. For example, compare the sentences (8) and (9).

(8)   Both x is greater than 1 or x is less than 1 and
           x is less than 0.
(9) x is greater than 1 or both x is less than 1 and
           x is less than 0.

Both sentences can be symbolized by letting

$$P = `x \text{ is greater than } 1\text{'}$$
$$Q = `x \text{ is less than } 1\text{'}$$
$$R = `x \text{ is less than } 0\text{'}$$

However, (8) is symbolized

(10) $(P \lor Q) \& R$

and (9) is symbolized

$$P \lor (Q \& R).$$

Put parentheses in the English sentences if they are needed to make the form clear. Note again that the parentheses enclose the molecular sentence which does *not* have the dominant connective. The dominant connective is *outside* the parentheses.

The exact and careful use of the parentheses in logic is quite important, for the sentence (P ∨ Q) & R is different from the sentence P ∨ (Q & R). Parentheses are required to indicate which connective dominates in each sentence.

EXERCISE 9

A. Each of the following symbolized sentences is a *conjunction*. This means, of course, that the major or dominant connective is 'and'. Put parentheses in where needed to show that the 'and' is dominant.

1. P ∨ Q & S
2. Q ∨ R & S
3. Q & R ∨ T
4. P ∨ R & Q
5. R & P ∨ T

B. Each of the following sentences is a *disjunction*. Put parentheses in where needed to show that in each case the major connective is the 'or'.

1. P ∨ Q & S
2. Q ∨ R & S
3. Q & R ∨ T
4. P & Q ∨ R
5. P ∨ Q & R

C. Each of the following sentences is listed as either a conjunction or a disjunction. Indicate the proper grouping of the atomic sentences by adding parentheses to show which connective dominates.

1. a disjunction	S ∨ T & R
2. a conjunction	T ∨ S & Q
3. a conjunction	T & S ∨ R
4. a disjunction	P ∨ Q & T
5. a disjunction	P & Q ∨ R

D. Symbolize the following sentences, indicating the grouping by parentheses when necessary.

1. Either Randy is President and Jim is Treasurer, or Bert is Treasurer.
2. Randy is President, and either Jim is Treasurer, or Bert is Treasurer.
3. Either Nicky is his brother and Sue is his sister or Larry is his brother.
4. Nicky is his brother and either Sue is his sister or Larry his brother.

5. Jeff is the captain or Ted is the captain, and Gary is the quarterback.

6. Both the answer is a prime number or Mary is wrong and Carol is wrong too.

E.   Symbolize the following mathematical sentences, choosing atomic letters to stand for the atomic mathematical sentences.

1. If x is less than 2, then x is equal to 1 or x is equal to 0.

2. If both x is less than three and x is greater than one then x equals two.

3. $y = 4$ and if $x < y$ then $x < 5$.

4. Either x is greater than five and x is less than seven or x is not equal to six.

5. If $x + 3 > 5$ and $y - 4 >$ then $y > 6$.

F.   Symbolize the five mathematical sentences of **E** above using the logical symbols for the connectives and mathematical symbols for the mathematical atomic sentences.

Now consider the sentence

> If this square is black then that square is red and your king is on a red square.

To symbolize this molecular sentence we let

> **P** = 'This square is black'
> **Q** = 'That square is red'
> **R** = 'Your king is on a red square'.

The symbolized sentence is

$$P \rightarrow (Q \ \& \ R).$$

The sentence is a conditional sentence in which the *consequent* (the sentence following 'then') is a conjunction. The major connective is 'if . . . then . . .' .

How can we change our example so that the connective 'and' is the major connective? In English we may do this by inserting a comma.

> If this square is black then that square is red,
> and your king is on a red square.

We may, if we wish, avoid the comma by using the word 'both' as part of the major connective.

> Both if this square is black then that square is
> red and your king was on a red square.

To show that & is now the major connective in the symbolized sentence, we change the position of the parentheses,

$$(P \rightarrow Q) \And R.$$

EXERCISE 10

A. For each sentence below you are given the name of the type of molecular sentence that is intended. Add the necessary parentheses.

1.	a conditional	$P \rightarrow R \And S$
2.	a conditional	$P \rightarrow Q \lor R$
3.	a conditional	$P \And Q \rightarrow R$
4.	a conditional	$R \lor P \rightarrow Q$
5.	a conjunction	$P \rightarrow Q \And S$
6.	a conjunction	$R \And P \rightarrow Q$
7.	a disjunction	$R \lor Q \rightarrow T$
8.	a disjunction	$Q \rightarrow P \lor S$
9.	a disjunction	$P \rightarrow R \lor Q$
10.	a conditional	$P \rightarrow R \lor Q$
11.	a conjunction	$P \And Q \rightarrow T$
12.	a conditional	$P \And Q \rightarrow T$
13.	a disjunction	$P \lor T \rightarrow Q$
14.	a disjunction	$Q \rightarrow R \lor \neg S$
15.	a conditional	$Q \rightarrow R \lor \neg S$

B. Symbolize the following sentences, indicating the grouping by parentheses where necessary. For all sentences, let

> J = 'Jerry is in Room 1'
> C = 'He is in chemistry class'
> K = 'Kenny is in Room 3'.

1. If Jerry is in Room 1, then Kenny is in Room 3 and he is in chemistry class.
2. If either Kenny is in Room 3 or he is in chemistry class, then Jerry is in Room 1.

3. Either if Jerry is in Room 1 then he is in chemistry class, or Jerry is not in Room 1.
4. Either Kenny is in Room 3 or if Jerry is in Room 1 then he is in chemistry class.
5. Both if Kenny is in Room 3, then he is in chemistry class, and Jerry is not in Room 1.

The Negation of a Molecular Sentence.   There are cases in which we want to show the negation of an entire molecular sentence. For example, we may wish to negate or deny a disjunction in a sentence of the following kind:

> It is not the case that the book is either red or
> that it is green.

Suppose we symbolize this sentence by first letting **P** be the name of the first atomic sentence and **Q** be the name of the last atomic sentence. The disjunction itself is **P** $\vee$ **Q**. Next, the form of the negation of a symbolized sentence is

$$\neg(\quad).$$

Note the symbol denoting the negation. Remember that we may negate any sentence, whether atomic or molecular. Any sentence can be negated by first placing it in parentheses and then placing a negation symbol in front of the parentheses. In symbolizing a sentence we must be aware that the symbol for negation applies to the smallest complete sentence in front of which it appears.

Thus to negate the sentence **P** $\vee$ **Q** put it in parentheses with a negation symbol in front.

$$\neg(\textbf{P} \vee \textbf{Q}).$$

The grouping of the parentheses shows (1) that the negation applies to the entire sentence (here a disjunction) — not just the nearest atomic sentence — and (2) that the negation is the major connective. In this example, the connective 'not' dominates the connective 'or'.

We may find examples in which other types of molecular sentences are negated. Again we state that parentheses are needed to show that the whole molecular sentence rather than just one of its parts is being denied. Consider the sentence

> It is not the case that both Joe has a sister
> and he has a brother.

Here we mean to negate the complete conjunction. In other words, we wish to say that Joe does not have both a sister and a brother. In symbolizing this sentence, if we let **P** be the name of the first atomic sentence and **Q** be the name of the last atomic sentence, then we have

$$\neg(P \ \& \ Q).$$

Finally, let us consider the negation of a conditional,

> It is not the case that if you see a black cat
> then you will have bad luck.

Let

> **P** = 'You see a black cat'.
> **Q** = 'You will have bad luck'.

Symbolized, our example will read

$$\neg(P \rightarrow Q).$$

The grouping of the parentheses shows us clearly that we are negating the complete conditional sentence and not merely the antecedent, sentence **P**.

Perhaps the simplest explanation for grouping and the use of parentheses in logic is that a molecular sentence enclosed in parentheses takes the same role as that of an atomic sentence in relation to other connectives and other sentences that may be joined with it. It is treated as a single sentence. The dominant connective is outside the parentheses.

EXERCISE 11

A. In each of the following sentences one of the symbols $\lor$, $\rightarrow$, or & dominates. Therefore the sentences are disjunctions, conditionals, and conjunctions although they all begin with negations. Suppose it was intended that the beginning negations dominate, making all of the sentences negations. With no changes except that of adding parentheses, make each sentence into a negation.

1. $\neg P \lor R$ 4. $\neg P \rightarrow Q$
2. $\neg R \rightarrow S$ 5. $\neg R \lor S$
3. $\neg P \ \& \ T$ 6. $\neg\neg Q \ \& \ \neg S$

B. Give the negation of each of the following sentences by adding negation symbols — and parentheses as needed.

1. S
2. P ∨ T
3. S & ¬T
4. P → R
5. Q & R
6. ¬R

7. T → ¬S
8. ¬N ∨ M
9. ¬Q → ¬T
10. ¬S & P
11. P ∨ ¬S
12. ¬Q

C. For each sentence below you are given the name of the type of molecular sentence that is intended. Add the necessary parentheses.

1. a negation   ¬P → R
2. a conditional   ¬P → R
3. a conjunction   ¬P & ¬R
4. a negation   ¬R & T
5. a conditional   ¬P → ¬Q
6. a negation   ¬P → ¬Q
7. a disjunction   ¬Q ∨ ¬R
8. a negation   ¬T ∨ S
9. a conjunction   ¬S & ¬Q
10. a negation   ¬R → S

D. Symbolize the following sentences, indicating the grouping by parentheses.

Let

P = 'This is Tuesday'
Q = 'It happened on Monday'.

1. Either this is not Tuesday or it did not happen on Monday.
2. It is not the case that if it happened on Monday, then this is Tuesday.
3. It is not the case that this is either Tuesday or that it happened on Monday.
4. It did not happen on Monday and this is Tuesday.
5. It is not the case that both this is Tuesday and it happened on Monday.
6. If it did not happen on Monday then this is not Tuesday.
7. It is not the case that if this is Tuesday then it happened on Monday.
8. Either this is not Tuesday or it happened on Monday.

9. This is not Tuesday and it happened on Monday.
10. It is not the case that both it happened on Monday and this is Tuesday.

E.   Symbolize the following sentences in exactly the way the example indicates.

1. Either Jack is the smallest and Tim is the tallest or Tim is the shortest and Jack is the largest.

Example: Let **P** = 'Jack is the smallest'
                    Q = 'Tim is the tallest'
                    R = 'Tim is the shortest'
                    S = 'Jack is the largest'.

$$(P \ \& \ Q) \lor (R \ \& \ S).$$

2. If an organic substance decays, then its compounds return to the soil and the soil is fertilized.
3. Either I am wrong, or question number one is true and question number two is false.
4. Both I am wrong or question number one is true, and question number two is false.
5. Either I am wrong and question number one is true or question number two is false.
6. It is not the case that both Jean is his sister and Jan is his sister.
7. Jean is not his sister and Jan is his sister.
8. If we know the period of the moon's motion and we know the distance from earth to moon, then we can compute the moon's centripetal acceleration.
9. Either his homework is finished, or if it is not finished then he must work on it tonight.
10. Not all countries of Africa have a hot and steaming climate and not all equatorial Africa is a land of thick and luxuriant vegetation.
11. If it is ten o'clock then the General Assembly session has begun, and that clock says ten o'clock now.
12. It is not the case that either very distant stars show a parallax or they appear as disks on a telescope.
13. If this rock is not hard, then it is not composed of quartz crystals.

14. If it is after five, then the door is locked and I do not have the key.

15. If it is after five then the door is locked, and furthermore, I do not have the key.

F.   For each of the following mathematical sentences you are given the type of molecular sentence it is intended to be. Add the necessary parentheses.

1. a conditional $x=0 \quad \vee \quad x=1 \quad \rightarrow \quad y=2$
2. a disjunction $x=0 \quad \vee \quad x\neq0 \quad \& \quad y=z$
3. a conjunction $x=1 \quad \vee \quad x\neq1 \quad \& \quad y\neq3$
4. a conditional $x=y \quad \rightarrow \quad y\neq z \quad \& \quad y>5$
5. a conjunction $x=y \quad \vee \quad x=z \quad \& \quad y>3$
6. a conditional $x=y \quad \& \quad y=z \quad \rightarrow \quad x=z$
7. a conditional $x>y \quad \& \quad y>z \quad \rightarrow \quad x>z$

▶ 1.7 *Elimination of Some Parentheses*

By adopting some simple rules about the *strength* of connectives, we may eliminate some of the parentheses in symbolized sentences.

● RULE 1

The → is stronger than the other three connectives.

Using Rule 1, in place of

$$(P \ \& \ Q) \rightarrow R$$

we may write simply

$$P \ \& \ Q \rightarrow R$$

Also, in place of

$$P \rightarrow (Q \vee R)$$

we may write

$$P \rightarrow Q \vee R.$$

On the other hand, if we have

$$(P \rightarrow Q) \vee R,$$

we cannot eliminate the parentheses, for we must use them to show that ∨ is the major connective. Also if a sentence has two conditional symbols, we must use parentheses to show which is dominant. Thus the sentence

$$A \rightarrow (B \rightarrow C)$$

has a different meaning from the sentence

$$(A \rightarrow B) \rightarrow C.$$

The second rule is so natural that we have already used it without explicit statement.

● RULE 2

The negation sign ¬ is weaker than any of the other three connectives.

Using Rule 2, in place of

$$(\neg P) \,\&\, Q$$

we write

$$\neg P \,\&\, Q,$$

or, in place of

$$P \lor (\neg Q)$$

we write

$$P \lor \neg Q,$$

or in place of

$$(\neg P) \rightarrow (\neg Q)$$

we write

$$\neg P \rightarrow \neg Q.$$

But the parentheses are necessary in

$$\neg (P \,\&\, Q).$$

Finally, we let & and ∨ have equal strength, so when both occur in a sentence, parentheses must always be used to indicate which is the major connective. Thus the meaning of

$$P \lor Q \,\&\, R$$

is not clear. But

$$(P \lor Q) \,\&\, R$$

is a conjunction, and

$$P \lor (Q \,\&\, R)$$

is a disjunction.

EXERCISE 1 2

A. For each of the following sentences you are given the type of molecular sentence that is intended. Using the two rules of strength, add parentheses *only* where necessary.

1. a conditional	P → Q ∨ R
2. a disjunction	P ∨ Q & R
3. a conjunction	R → S & T
4. a negation	¬R & S
5. a conditional	P ∨ Q → ¬R
6. a negation	¬P → Q
7. a conjunction	A & B → C
8. a disjunction	M → N ∨ P
9. a negation	¬P ∨ ¬Q
10. a conjunction	¬A ∨ ¬B & ¬C

B. For each of the following mathematical sentences you are given the type of molecular sentence that is intended. Using the two rules of strength, add parentheses *only* where necessary.

1. a conjunction	$x \neq 0 \quad \vee \quad x > y \quad \& \quad y = z$
2. a conditional	$x = 0 \quad \rightarrow \quad x > y \quad \& \quad y \neq z$
3. a disjunction	$x = 0 \quad \vee \quad x \neq 0 \quad \& \quad y = z$
4. a conditional	$x > y \quad \& \quad y > z \quad \rightarrow \quad x > z$
5. a disjunction	$x = 0 \quad \vee \quad x > 0 \quad \rightarrow \quad y = 0$
6. a conjunction	$x = y \quad \& \quad y = z \quad \vee \quad x = z$
7. a conditional	$x = y \quad \& \quad y = z \quad \rightarrow \quad x = z$
8. a conjunction	$x = y \quad \vee \quad x = z \quad \& \quad y \neq z$

C. Symbolize the sentences of Exercise 11, Section E, using parentheses *only* where necessary.

▶ 1.8 *Summary*

In order to symbolize sentences in logic, we must be able to recognize the logical parts of those sentences. A molecular sentence is made up of at least one atomic sentence plus a connective. An atomic sentence is one that does not have a connective. 'Sentential connective' (or simply 'connective') is the name in logic for such words as 'both . . .

and . . .', 'either . . . or . . .', 'if . . . then . . .' and 'not' that are used to make molecular sentences out of atomic sentences.

Of the four connectives listed, 'and', 'or' and 'if . . . then . . .' connect or control two sentences at a time, while the connective 'not' controls just one. A molecular sentence formed by using the connective 'and' is a conjunction, a molecular sentence formed by using the connective 'or' is a disjunction, a molecular sentence formed by using the connective 'not' is a negation, and a molecular sentence formed by using the connective 'if . . . then . . .' is a conditional sentence.

It is convenient in logic to use symbols for sentences and for connectives. For atomic sentences we use capital letters such as 'P', 'Q', 'R', 'S', and so forth. Since the connective determines the form of a sentence in logic, we can replace any atomic sentence by any other and the form will remain the same. For example, in sentence P & Q we can substitute the names of any English sentences for P and for Q. The symbols used for the connectives, on the other hand, always remain the same. They are: & for conjunction, ∨ for disjunction, ¬ for negation, and → for conditional.

In sentences having more than one connective we need to indicate the grouping, for a different grouping can have a different meaning. In English, grouping is shown according to where certain words are placed, or by punctuation. In logic, grouping is shown by parentheses. The conjunction (P ∨ Q) & R has a different meaning from the disjunction P ∨ (Q & R), even though the atomic sentences and connectives are the same. We need parentheses to show when a connective dominates the sentence, if it is not the strongest connective in that sentence. 'Not' is weakest; next are 'and' and 'or', which are equal in strength; and 'if . . . then . . .' is strongest. Any connective may dominate, however, if the parentheses show it.

With these symbols as tools we are now prepared to express clearly and precisely the logical meaning of all but a few sentences that are possible within the part of elementary formal logic known as sentential logic.

EXERCISE 13

Review Exercise

A.   Put an 'A' after each sentence that is atomic and an 'M' after each sentence that is molecular. After each molecular sentence write the sentential connective used in the particular sentence.

1. Weather is the atmosphere's condition at a particular time and climate is the average of weather conditions over a longer period of time.
2. The bacteria in the water are either destroyed by boiling the water or they are destroyed by chlorination.
3. This book has more pages than the other one.
4. If the defendant is ruled against, then he will appeal the case.
5. He recognized the work to be that of a Nineteenth Century English poet.
6. The war cannot be fully explained in terms of one cause.
7. An element has physical properties and it has chemical properties.
8. We are able to do all of the exercises on this page.
9. We are not able to do all of the exercises on this page.
10. If two or more elements join chemically to make a new substance, then the product is called a compound.
11. Molecular sentences contain connectives.
12. This problem is not correct.
13. Karen is a sophomore and her brother is a senior.
14. We cannot finish the report today.
15. We shall need help or we shall take two days to complete the report.

B.   Write four sentences which have connectives. Use a *different* connective in each sentence.

C.   Write four atomic sentences.

D.   Symbolize the following sentences, indicating what atomic sentence each capital letter symbol represents.

1. If it is after five, then the meeting is over.
2. Either my watch is wrong or we are going to be late.
3. If the plant cell does not have chlorophyll, then it cannot make food.
4. Sandstone is produced by layers of sand hardening and limestone is produced from the shells of small animals in the sea.
5. If the tribe was nomadic, then it did not build permanent shelters.

E.   Symbolize the following sentences, using the given symbols for the atomic sentences.

Let

> P = 'Bob is too late'
> Q = 'Jane is too early'
> R = 'Mr. Jones is angry'.

1. If Bob is too late and Jane is too early, then Mr. Jones is angry.
2. If either Bob is too late or Jane is too early, then Mr. Jones is angry.
3. If Bob is too late and Jane is not too early, then Mr. Jones is not angry.
4. If Mr. Jones is angry, then Bob is too late or Jane is not too early.
5. Mr. Jones is angry, and Bob is too late and Jane is too early.
6. If Mr. Jones is not angry, then Bob is not too late.
7. Either Bob is too late or Jane is too early.
8. If Jane is not too early or Bob is too late, then Mr. Jones is angry.
9. Mr. Jones is angry and either Bob is too late or Jane is too early.
10. Jane is too early, and if Bob is too late, then Mr. Jones is angry.
11. It is not the case that both Bob is too late and Jane is too early.
12. If Bob is not too late and Jane is too early, then Mr. Jones is not angry.

F.   Complete the translation of the following molecular sentences into logical symbols by replacing the words for the connectives by the proper symbol.

1. If P then Q
2. Either P or Q
3. If either P or Q then not R
4. Either not P or not Q
5. Either P and Q, or R and S
6. It is not the case that both P and Q
7. It is not the case that either P or Q
8. If not P, then not Q and R
9. It is not the case that if P then Q
10. It is not the case that both P and not P

11. **P** and either **Q** or **R**
12. Either **P** and **Q** or **R**
13. **P** and if **Q** then not **R**

G. Match each of the words on the left with the examples or definitions in the list at the right.

1. disjunction
2. negation
3. conditional sentence
4. molecular sentence
5. antecedent
6. consequent
7. conjunction
8. atomic sentence

(a) **P** → **Q**
(b) ¬(**P** & **Q**)
(c) **P** ∨ **Q**
(d) **Q** in the sentence **P** → **Q**
(e) ¬**P**
(f) **P** in the sentence **P** → **Q**
(g) **P** & **Q**
(h) ¬**P** ∨ ¬**Q**
(i) Any sentence with a connective
(j) Any sentence without a connective

H. Symbolize the following mathematical sentences by choosing capital letters to stand for the atomic mathematical sentences and tell what each atomic sentence is.

1. x is greater than five.
2. Four is not an odd number.
3. x is equal to three or x is greater than six.
4. It is not the case that if x is an odd number then x is divisible by two.
5. If x plus four is seven and y plus x is eight then y is five.
6. If x is less than five or greater than seven then it is not equal to six.

I. Symbolize mathematical sentences (3), (5), and (6) of Part **H** by using the logical symbols for the connectives and standard mathematical symbols for the atomic sentences.

J. Translate the following logical sentences (formulas) into English. First choose an English atomic sentence for each atomic letter, tell what English sentence you chose, and then write out the sentence completely in English.

1. ¬**S**
2. **P** ∨ ¬**Q**

3. $\neg(R \rightarrow S)$
4. $x < 5 \quad \rightarrow \quad \neg(x > 6)$
5. $x + 3 < 5 \quad \& \quad \neg(x = 0) \quad \rightarrow \quad x = 1$
6. $P \ \& \ \neg Q \rightarrow R$

REVIEW TEST

I. *Symbolizing Everyday Language*

Symbolize the following sentences, stating clearly what the capital letter symbols you choose represent. For the mathematical sentences use standard mathematical symbols.

    a. If the book costs more than two dollars, then Jane will not buy it.

    b. Either this is Bob's house or the directions given us are not correct.

    c. Rising air expands and it becomes colder.

    d. If x is less than three, then it is less than four.

    e. If x is not equal to five, then either it is less than five or it is greater than five.

II. *Symbolizing with Given Symbols*

Using the symbols given, symbolize the following sentences. (You will not need to write the English sentences here.)

    Let

$$P = \text{'John is too early'}$$
$$Q = \text{'Mary is too late'}$$
$$R = \text{'Mr. Smith is angry'}$$

    a. If John is too early or Mary is too late, then Mr. Smith is angry.

    b. If Mary is too late, then John is not too early.

    c. Either Mr. Smith is angry or Mary is not too late.

    d. Mary is too late and John is too early, and Mr. Smith is angry.

    e. If Mr. Smith is not angry, then John is not too early and Mary is not too late.

    f. Either Mary is not too late or John is too early.

    g. If Mary is not too late and John is not too early, then Mr. Smith is not angry.

III. *Definitions*

Complete the following sentences by choosing the word from the list below which you think is defined by the sentence.

    a. The molecular sentence that uses the connective 'and' is a _____.

    b. The molecular sentence that uses the connective 'not' is a _____.

    c. The combination of one or more atomic sentences with a sentential connective is called a _____.

    d. In logic, a complete sentence that has no sentential connective is called _____.

    e. The molecular sentence that uses the connective 'if . . . then . . .' is called a _____.

    f. The sentence before the connective in a conditional sentence is called the _____.

    g. The sentence after the connective in a conditional sentence is called the _____.

    h. The molecular sentence that uses the connective 'or' is a _____.

antecedent	conjunction
atomic	consequent
molecular sentence	disjunction
conditional	negation

IV. *Use of Parentheses*

For some of the following symbolized sentences, parentheses are needed to make them the kind of molecular sentence indicated at the left. Put parentheses in the proper places if needed.

a.	a conjunction	P ∨ Q & R
b.	a negation	¬P & Q
c.	a conjunction	¬P & Q
d.	a conditional	P & Q → R
e.	a negation	¬P ∨ ¬R
f.	a disjunction	P → Q ∨ R
g.	a conditional	¬P → ¬R
h.	a disjunction	P ∨ Q & R

i. a negation $\neg P \to Q$

j. a conjunction $P \;\&\; Q \to R$

V. *Symbolizing Sentences with Parentheses*

Circle the major or dominant connective in the following sentences. Then show the sentence as it should appear in logical symbols and add parentheses where needed.

a. It is not the case that either Jack is tallest or that Jill is tallest.

b. Tom is not our representative and Larry is not our captain.

c. Either 'beta' comes before 'gamma' and 'eta' comes before 'theta', or I do not know Greek.

d. Jim is leaving now and either I will go with him or Terry will go with him.

e. If the dance starts at eight, then we will be early and Mike will be late.

CHAPTER TWO
LOGICAL INFERENCE

▶ 2.1 *Introduction*

In Chapter One we learned to separate sentences into their logical parts. In this way we learned something about the logical form of sentences. The idea of *form* is illustrated by something you already know. The sentence **P** → **Q** is the same in logical form no matter what English sentences **P** and **Q** may stand for. The sentential connectives determine the form of a sentence.

Knowing the forms of sentences and having the tools of symbolization at our command, we can now go on to an important concern of formal logic: inference and deduction. The *rules of inference* controlling the sentential connectives turn out to be very simple. We can learn these rules and how to use them just as if we were learning to play a game. The game is played with sentences, or logical formulas, a name we shall give to our symbolized sentences. We begin with sets of formulas which are called *premises*. The object of the game is to use the rules of inference in such a way that we are led to other formulas, which are called *conclusions*. To go logically from premises to conclusion is to perform a *deduction*. The conclusion you reach may be said to *follow logically from* the premises if every move you made to reach that conclusion is permitted by a rule. The idea of inference may be summed up in this way: *from premises that are true we reach only conclusions that are true.* In other words, if the premises are true, then the conclusions that are *derived* logically from them *must* be true.

We usually learn a new game by example. Let us look at some examples of inference before going on to the formal rules. Suppose we have two premises, the formula **P** → **Q** and the formula **P**. We say that these premises are given; that is, we start by saying that we are given **P** and that we are given **P** → **Q**. Can you infer a conclusion from those two sentences? In other words, can you think of another sentence that must be true if the premises are true? The conclusion is clear if we

43

read the premises as

If P then Q, and P.

The first sentence tells us that if we have **P**, then we have **Q** and the second tells us that we do have **P**. The conclusion is that we have **Q**. The sentence **Q** follows logically from the premises, **P** and **P → Q**.

Now let us look at an inference of the same form but whose content is supplied by ordinary language. The first premise is

If it is raining, then the sky must be cloudy.

The second premise is

It is raining.

What conclusion can we infer from the two premises?

The answer is the conclusion 'The sky must be cloudy'. This conclusion may be inferred logically from the given premises. We may say that it follows logically from the premises. We will go on now to discuss the particular rule of inference which permits us to derive that conclusion from the premises.

▶ 2.2 *Rules of Inference and Proof*

Modus Ponendo Ponens. The rule of inference applied in the preceding example was one which has the Latin name, *modus ponendo ponens.* Let us consider some examples of the use of this rule in deriving conclusions from premises.

Premise 1.	If he is on the football team, then he is at the practice field now.
Premise 2.	He is on the football team.
Conclusion.	He is at the practice field now.

Another example of the use of *modus ponendo ponens* is the following:

Premise 1.	If it is not cold, then the lake will not be frozen.
Premise 2.	It is not cold.
Conclusion.	The lake will not be frozen.

In symbolic terms, we illustrate the first example thus:

Let

> **P** = 'He is on the football team'
> **Q** = 'He is at the practice field now'.

Then

Premise 1.	**P → Q**
Premise 2.	**P**
	————
Conclusion.	**Q**

The rule of inference called *modus ponendo ponens* allows us to *prove* **Q** from **P → Q** and **P**.

The second example is symbolized in the following way, where **P** is the sentence 'It is cold' and **Q** is the sentence 'The lake will be frozen'.

> ¬**P** → ¬**Q**
> ¬**P**
> ————
> ¬**Q**

In each of the examples the rule *modus ponendo ponens* allows us to go from the two premises to the conclusion. To say that the conclusion follows logically from the premises is to say that whenever the premises are true, then the conclusion is also true. The rule of inference we have learned tells us that if we have any two sentences of the form **P → Q** and **P**, we can derive the conclusion **Q**.

Remember that the rule applies to the *form* of the sentences. It means that whenever we are given a *conditional sentence* and also precisely *the antecedent of that conditional*, then precisely the *consequent* follows. The same rule applies whether the antecedent is an atomic sentence or a molecular sentence and whether the consequent is an atomic sentence or a molecular sentence. In the conditional sentence above the antecedent and consequent are molecular sentences. The second premise asserts or states precisely the antecedent, which is ¬**P**. Therefore, precisely the consequent, which is ¬**Q**, follows by the rule of *modus ponendo ponens*. The following examples all apply *modus ponendo ponens*. Notice that both molecular and atomic sentences are used as antecedents and consequents.

a. R → S   b. P   c. P & Q → R
   R   P → ¬Q   P & Q
  ———   ————   ————
   S   ¬Q   R

        d. ¬P → Q   e. P → Q & R
           ¬P   P
          ————   ————
           Q   Q & R

Notice in the second example that the conditional was placed second and P, which is precisely its antecedent, was placed first. When *modus ponendo ponens* or any of the other rules are applied to derive a conclusion from two or more sentences the order of those sentences makes no difference.

Recall that a conditional may be written (P) → (Q). With the parentheses, *modus ponendo ponens* may be written

$$(P) \rightarrow (Q)$$
$$(P)$$
$$————$$
$$(Q)$$

If it helps, put in the parentheses when either the antecedent or consequent are themselves molecular sentences, as in the last three examples above or in the following one.

¬P ∨ R → S & ¬Q   (¬P ∨ R) → (S & ¬Q)
¬P ∨ R   (¬P ∨ R)
————————   ——————————
S & ¬Q   (S & ¬Q)

The name *modus ponendo ponens* may be explained in this way. This rule of inference is the method (*modus*) that affirms (*ponens*) the consequent by affirming (*ponendo*) the antecedent. It is also called the *Law of Detachment*, because if we are given the antecedent, we may *detach* the consequent and assert it.

EXERCISE 1

A.   What conclusion can you derive from each of the following sets of premises? In other words, what sentence logically follows from the premises?

1. If you are on the West Coast, then you are on Pacific Standard Time. You are on the West Coast.

2. If we do not leave at once, then we will not make it to the plane. We do not leave at once.
3. If this plant does not grow, then either it needs more water or it needs better soil. This plant does not grow.
4. It is five o'clock. If it is five o'clock, then the office is closed.
5. If I live in the capitol of the United States, then I do not live in one of the fifty states. I live in the capitol of the United States.

B.   Derive a conclusion from each of the following sets of premises, using *modus ponendo ponens*. Put your conclusion on line (3).

1. (1) P ∨ Q → R
   (2) P ∨ Q
   (3)

2. (1) ¬P → ¬R
   (2) ¬P
   (3)

3. (1) ¬P
   (2) ¬P → Q
   (3)

4. (1) P → Q & R
   (2) P
   (3)

5. (1) P → Q ∨ R
   (2) P
   (3)

6. (1) ¬R
   (2) ¬R → Q & P
   (3)

C.   Write a 'C' for each example of a correct conclusion permitted by *modus ponendo ponens*. Write an 'I' for incorrect conclusion.

1. Premises: S and S → T; conclusion: T
2. Premises: T → V and T; conclusion: V
3. Premises: P → Q and Q; conclusion: P
4. Premises: S and R → S; conclusion: R
5. Premises: R and R → S; conclusion: S

D.   Use *modus ponendo ponens* to derive a conclusion from each of the following sets of premises.

1. If $x \neq 0$ then $x+y>1$. $x \neq 0$.
2. If $x+y=z$ then $y+x=z$. $x+y=z$.
3. If x is a number and y is a number, then $x+y$ is a number. x is a number and y is a number.
4. If $x>y$ and $y>z$, then $x>y$.   Both $x>y$ and $y>z$.
5. Both $x=y$ and $y=z$.   If $x=y$ and $y=z$, then $x=z$.

Proofs. When we use a rule of inference to move from a set of sentences to another sentence we are *proving* that the last sentence follows logically from the others. This may be said in many ways. We can say that we are *deriving* the conclusion from the premises, that the conclusion is *inferred* from or *implied* by the premises, that we *deduce* the conclusion from the premises, and so forth. All of these words or expressions mean the same thing. We have certain given sentences. If a rule of inference permits us to move to another sentence, then that sentence is a logical conclusion from the given sentences.

In the last section we were shown several short proofs. Using *modus ponendo ponens* as our rule, we proved a conclusion from sets of premises. For example, from $R \rightarrow S$ and R we proved S. We could set the proof out very clearly by

$$(1) \; R \rightarrow S \qquad P$$
$$(2) \; R \qquad\qquad P$$
$$(3) \; S \qquad\qquad PP$$

Each line in our proof is numbered. After the symbolized sentences, we show how we obtained each sentence. For the premises, which are given, we write 'P', which simply means that the sentence is a premise. We justify lines that are premises by the *Rule of Premises*, P. We start with these premises. We derived line (3) by *modus ponendo ponens*. This is indicated by the abbreviation for the rule, PP, after the sentence.

EXERCISE 2

A. Below, you are given sets of premises. Derive a conclusion from each set and show, by abbreviations for Rule of Premises, P, or *modus ponendo ponens*, PP, how you obtained each of the three lines.
Example:

$$(1) \; \neg P \rightarrow S \qquad P$$
$$(2) \; \neg P \qquad\qquad P$$
$$(3) \; S \qquad\qquad PP$$

1. (1) $\neg A \rightarrow \neg B$
   (2) $\neg A$
   (3)

2. (1) M
   (2) $M \rightarrow N$
   (3)

3. (1) R
   (2) $R \rightarrow \neg T \lor Q$
   (3)

4. (1) $\neg B \rightarrow \neg D \mathbin{\&} A$
   (2) $\neg B$
   (3)

B.   Symbolize each of the sets of premises under **A** in Exercise 1. Then show a proof as in Section **A**, this exercise, numbering each line and showing by the abbreviations, either P for premise or PP for *modus ponendo ponens*, how you justify each line.

C.   Symbolize the mathematical sentences under Section **D** of Exercise 1. Then show a proof as in Section **A**, this exercise.

Two-Step Proofs.   Sometimes we cannot go directly from premises to the conclusion we want in just a single step. But this does not prevent us from reaching that conclusion. Each time we derive a sentence by a rule, then that sentence can be used along with the premises to derive another sentence. Let us consider an example in which we have three premises.

$$(1)\ A \rightarrow B \qquad P$$
$$(2)\ B \rightarrow C \qquad P$$
$$(3)\ A \qquad P$$

We wish to prove sentence **C**. Two steps are needed to reach **C**, each permitted by *modus ponendo ponens*, PP. These two steps are lines (4) and (5) below.

$$(1)\ A \rightarrow B \qquad P$$
$$(2)\ B \rightarrow C \qquad P$$
$$(3)\ A \qquad P$$
$$(4)\ B \qquad PP\ 1,\ 3$$
$$(5)\ C \qquad PP\ 2,\ 4$$

Notice these things about the proof. Each line is numbered, whether it is a premise or a derived line. Each line is justified as either a premise (signified by P) or as derived by a rule of inference (signified by the abbreviation for the rule, PP). In addition, after the abbreviation for the rule for all *derived* lines we put the numbers of the lines from which that line was derived. For example, in line (4) the code 'PP 1, 3' means that **B** was derived by *modus ponendo ponens* from lines (1) and (3). Similarly in line (5), we see that **C** was derived by rule PP from lines (2) and (4). Go over each step to see how the rule was applied to certain lines. Notice also that we can use a line that we have derived, together with other lines, to derive a new line. Any line that can be justified either as a premise or by use of a rule can be used in other steps in the proof.

Before trying some short proofs, consider one more example. Suppose we are given the following premises and asked to prove **R**.

(1)	S → ¬T	P
(2)	S	P
(3)	¬T → R	P
(4)	¬T	PP 1, 2
(5)	R	PP 3, 4

We use *modus ponendo ponens* to derive one line (4) so that we can apply *modus ponendo ponens* to that line and some other line (3) to derive the conclusion (5). We take one step (permitted by a rule) so that we can take another step by using the sentence we derived.

EXERCISE 3

A.   In each of the exercises below you are asked to prove that a certain sentence follows logically from the given premises. Derive the conclusion, writing the abbreviation for the rule that permits a line and, for derived lines, show the number of each line that was used in applying the rule.

1. Prove: ¬T

(1)	R → ¬T	P
(2)	S → R	P
(3)	S	P
(4)		
(5)		

3. Prove: C

(1)	A → B & D	P
(2)	B & D → C	P
(3)	A	P
(4)		
(5)		

2. Prove: G

(1)	¬H → ¬J	P
(2)	¬H	P
(3)	¬J → G	P
(4)		
(5)		

4. Prove: M ∨ N

(1)	¬J → M ∨ N	P
(2)	F ∨ G → ¬J	P
(3)	F ∨ G	P
(4)		
(5)		

5. Prove: ¬S

(1)	T	P
(2)	T → ¬Q	P
(3)	¬Q → ¬S	P
(4)		
(5)		

B. Symbolize each of the following sets of premises and prove that the conclusion (the sentence beginning 'Therefore . . .') follows logically. This should be done in just the way the proofs on page 50 were shown.

1. If 2 is greater than 1, then 3 is greater than 1.
   If 3 is greater than 1, then 3 is greater than 0.
   2 is greater than 1.
   Therefore, 3 is greater than 0.

2. $x+1=2$.*
   If $x+1=2$ then $y+1=2$.
   If $y+1=2$ then $x=y$.
   Therefore, $x=y$.

3. If $x+0=y$ then $x=y$.
   $x+0=y$.
   If $x=y$ then $x+2=y+2$.
   Therefore, $x+2=y+2$.

4. If $x>y$ and $y>z$ then $x>z$.
   $x>y$ and $y>z$.
   If $x>z$ then $x>10$.
   Therefore, $x>10$.

5. If $x=y$ and $y=z$ then $x=z$.
   If $x=z$ then $z=x$.
   $x=y$ and $y=z$.
   Therefore, $z=x$.

6. If the moist air rises, then it will cool.
   If it cools, then clouds will form.
   The moist air rises.
   Therefore, clouds will form.

C. There is no limit to the number of times the rule *modus ponendo ponens* may be used in one proof. The following exercises require more than two applications. Derive the conclusion you are asked to prove, stating the rule used to derive each line and indicate the lines that were used in applying the rule.

---

* When mathematical symbols are used to express an atomic sentence, we do not need to use a capital letter to symbolize the atomic sentence, for we shall use mathematical as well as logical symbols. For example, in Exercise 2 under **B** the premises may be written

$$x+1=2$$
$$x+1=2 \quad \rightarrow \quad y+1=2$$
$$y+1=2 \quad \rightarrow \quad x=y$$

1. Prove: ¬N

    (1) R → ¬S   P

    (2) R   P

    (3) ¬S → Q   P

    (4) Q → ¬N   P

2. Prove: B

    (1) ¬G → E   P

    (2) E → K   P

    (3) ¬G   P

    (4) K → ¬L   P

    (5) ¬L → M   P

    (6) M → B   P

3. Prove: R ∨ S

    (1) C ∨ D   P

    (2) C ∨ D → ¬F   P

    (3) ¬F → A & ¬B   P

    (4) A & ¬B → R ∨ S   P

Double Negation. The *Rule of Double Negation* is a simple rule that allows us to go from a single premise to a conclusion. As an example, we begin with the negation of a negation, as the name 'double negation' suggests. Look at the sentence,

It is not the case that Ann is not a student.

What can you conclude from this premise? Obviously we can say

Ann is a student.

The Rule of Double Negation works the other way also. For example, from the sentence,

John takes the bus to school

we can conclude the negation of its negation.

It is not the case that John does not take
the bus to school.

Thus the Rule of Double Negation has two forms in symbolic terms.

$$\frac{(P)}{\neg\neg(P)} \quad \text{and} \quad \frac{\neg\neg(P)}{(P)}$$

The abbreviation for the rule is DN.

Some examples of the use of double negation in proving that a conclusion follows logically from a premise are now shown.

a. (1) **R**   P
    (2) ¬¬**R**   DN 1

b. (1) ¬¬**A**   P
    (2) **A**   DN 1

c. (1) ¬¬(**P** & **Q**)   P
    (2) **P** & **Q**   DN 1

Now that we have learned two rules of inference we can do short proofs that require the use of both. Consider the example below in which both *modus ponendo ponens*, PP, and double negation, DN, are used to reach the conclusion.

(1) **P** → **Q**   P
(2) **P**   P
(3) **Q**   PP 1, 2
(4) ¬¬**Q**   DN 3

There are two premises and two derived lines in the proof. Line (3) is derived from lines (1) and (2) by *modus ponendo ponens*. Line (4) is derived from line (3) by the rule of double negation.

EXERCISE 4

A.   What conclusions can you infer from each of the following sentences by double negation?

1. All mammals are warm-blooded animals.
2. It is not the case that the nucleus of the atom is not positively charged.
3. Granite is a type of igneous rock.
4. In the United States presidential elections are held every four years.
5. It is not the case that one fifth is not equal to twenty per cent.

B.   For each of the following sets of premises derive a conclusion, if possible, by *modus ponendo ponens*. If the rule *modus ponendo ponens* cannot be applied to the premises, then indicate this by writing 'not PP'.

1. (1) **P** & **Q** → **R**
    (2) **R**

2. (1) **Q** → **R** ∨ **S**
    (2) **Q**

3. (1) ¬¬**R**
    (2) **Q** → ¬¬**R**

4. (1) **S**
    (2) **S** → ¬**P**

5. (1) **S** → **T** & **U**
    (2) **T** & **U**

6. (1) ¬¬**P** → **Q**
    (2) ¬¬**P**

C.   Put the letter T for each true statement. Put the letter F for each false statement.

1. From ¬¬R we may derive **R**.
2. From **S** we may derive ¬**S**.
3. From **P** → **Q** and **P** we may derive **Q**.
4. From **Q** we may derive ¬¬**Q**.
5. From **R** → **S** and **S** we may derive **R**.

D.   Prove that the conclusion follows logically from the given premises in each example below. Show the *complete proof* as in the examples given earlier. This means that you should number each line, show the abbreviation of the rule used, and the numbers of lines from which you derived each line in the proof.

1. Prove: ¬¬T
   (1) S → T   P
   (2) S   P
   (3)
   (4)

2. Prove: **B**
   (1) ¬A   P
   (2) ¬A → ¬¬B   P
   (3)
   (4)

3. Prove: **G**
   (1) H → ¬¬G   P
   (2) H   P
   (3)
   (4)

4. Prove: **P** ∨ **Q**
   (1) R → ¬¬(P ∨ Q) P
   (2) R   P
   (3)
   (4)

5. Prove: ¬¬N
   (1) M → ¬P   P
   (2) ¬P → N   P
   (3) M   P
   (4)
   (5)
   (6)

6. Prove: **Q**
   (1) J → K & M   P
   (2) J   P
   (3) K & M → ¬¬Q   P
   (4)
   (5)
   (6)

Modus Tollendo Tollens.   The rule of inference that has the Latin name *modus tollendo tollens* deals with conditionals. But in this case, by denying (tollendo) the consequent, we may deny (tollens) the antecedent of the conditional. The following derivation is an example of the use of *modus tollendo tollens.*

Premise 1.   If it shines by its own light, then the heavenly object is a star.

Premise 2.   The heavenly object is not a star.

Conclusion.   Therefore it does not shine by its own light.

We would symbolize the example in this way:

Let

$$P = \text{'It shines by its own light'}$$
$$Q = \text{'The heavenly object is a star'}.$$

$$P \rightarrow Q$$
$$\neg Q$$

Therefore $\neg P$

The abbreviation for *modus tollendo tollens* is TT.

When either the antecedent or the consequent is a molecular sentence, it may help to put in parentheses to make the form clear.

$$(P) \rightarrow (Q)$$
$$\neg(Q)$$

Therefore $\neg(P)$

The rule *modus tollendo tollens* permits us to go from two premises, (a) a conditional sentence and (b) a sentence denying the consequent, to a conclusion in which we deny the antecedent.

Let us look at another example that may help in making the previous statement clear. The conditional sentence is

If it is morning, then the sun will be in the east.

We deny its consequent

The sun is not in the east.

Thus we can deny its antecedent

Therefore, it is not morning.

The rule applies to all sets of premises of this *form*. The antecedent or the consequent may be either molecular sentences or atomic sentences. Look at the following examples, all of which use the rule *modus tollendo tollens*. Each has a conditional sentence for one premise; another premise denies the consequent.

a. (1) R → S   P   *b.* (1) Q & R → S   P
    (2) ¬S   P   (2) ¬S   P
    (3) ¬R   TT 1, 2   (3) ¬(Q & R)   TT 1, 2

               c. (1) P → ¬Q   P
                   (2) ¬¬Q   P
                   (3) ¬P   TT 1, 2

Notice that in the last example, we deny a negation by a double negation. To deny ¬Q we need ¬¬Q. A double negation is the negation of a negation.

Now let us look at an example of a proof that applies all three of the rules learned so far. We want to prove ¬¬R.

               (1) P → Q   P
               (2) ¬Q   P
               (3) ¬P → R   P
               (4) ¬P   TT 1, 2
               (5) R   PP 3, 4
               (6) ¬¬R   DN 5

Go over this example to make sure that you can follow every step that was taken. Here is another example using just two rules. We wish to prove A.

               (1) ¬A → ¬B   P
               (2) B   P
               (3) ¬¬B   DN 2
               (4) ¬¬A   TT 1, 3
               (5) A   DN 4

The use of double negation is important here. We need the negation of the consequent in the first premise in order to apply rule TT. The consequent is ¬B. The negation of that molecular sentence is shown by adding the symbol for 'not'. So we know that ¬¬B denies or negates ¬B. We do not have ¬¬B in the premises but we can derive it from B, the second premise. Notice that this was done in line (3). By using *modus tollendo tollens* we get the negation of the antecedent. The antecedent is ¬A so its negation is ¬¬A. Then all that remains is to apply rule DN again to derive A from ¬¬A.

EXERCISE 5

A. What conclusion can you derive from each of the following sets of premises using rule TT? State your conclusions in English sentences.

1. If light were simply a continuous wave motion, then brighter light would always cause electrons to escape with greater energy than would dimmer light. Brighter light does not always cause electrons to escape with greater energy than does dimmer light.

2. If one angle of this triangle is greater than 90 degrees, then the sum of the other two angles is less than 90 degrees. The sum of the other two angles is not less than 90 degrees.

3. If the lease is held to be valid, then the owner is responsible for the repairs. The owner is not responsible for the repairs.

4. If it rained last night, then the tracks are washed away. The tracks are not washed away.

5. Bob is not my brother. If Susan is my sister then Bob is my brother.

B. Derive a conclusion from each of the following sets of premises, using the rule *modus tollendo tollens*.

1. (1) $Q \to R$ P   4. (1) $Q \to \neg R$ P
   (2) $\neg R$ P   (2) $\neg\neg R$ P
   (3)   (3)

2. (1) $\neg P \to Q$ P   5. (1) $P \to Q$ & R P
   (2) $\neg Q$ P   (2) $\neg(Q$ & $R)$ P
   (3)   (3)

3. (1) $R \to S$ P   6. (1) $P \lor Q \to R$ P
   (2) $\neg S$ P   (2) $\neg R$
   (3)   (3)

C. Prove that the conclusions follow from the given premises. Show a complete proof.

1. Prove: **C**   2. Prove: **F**
   (1) $\neg B$ P   (1) $G \to H$ P
   (2) $A \to B$ P   (2) $\neg G \to \neg\neg F$ P
   (3) $\neg A \to C$ P   (3) $\neg H$ P

3. Prove: R & S   4. Prove: **E**

   (1) **P → ¬Q**   P   (1) **F**   P

   (2) **Q**   P   (2) **¬E → ¬F**   P

   (3) **¬P → R & S**   P

            5. Prove: **¬S**

              (1) **S → ¬R**   P

              (2) **R**   P

More on Negation. The Rule of Double Negation is used frequently with *modus tollendo tollens,* and with other rules shortly to be introduced. Since the use of the Rule of Double Negation in conjunction with TT always assumes essentially the same form, we may shorten derivations by the introduction of an extended definition of negation:

> **P** *is the negation of* **¬P**.

We already know that **¬P** is the negation of **P**. We can use the Rule of Double Negation to get this extension of a definition of negation. Given **¬P**, we know that its negation is **¬¬P**, but by the rule of double negation, we obtain the equivalent sentence **P**. This rule does not permit us to prove anything we could not prove without it.

    Recognition that **P** is the negation of **¬P** simplifies derivations in the way illustrated by

          (1) **A → ¬B**   P

          (2) **B**   P

          (3) **¬A**   TT 1, 2

From the two premises we get the negation of **A** at once by TT. The negation of **A** is **¬A**. This is because we accept **B** as the negation of **¬B**. Without this extended definition of negation, the derivation would need an extra line involving the Rule of Double Negation.

          (1) **A → ¬B**   P

          (2) **B**   P

          (3) **¬¬B**   DN 2

          (4) **¬A**   TT 1, 3

Notice that the effect of recognizing as **P** the negation of ¬**P** is to extend TT to the following logical form:

$$P \rightarrow \neg Q$$
$$Q$$

$$\overline{}$$

$$\neg P$$

A similar extension of TT concerns the antecedent of the conditional premise:

$$\neg P \rightarrow Q$$
$$\neg Q$$

$$\overline{}$$

$$P$$

This extension is used in the following example.

(1) ¬**A** → **B**   P
(2) ¬**B**   P
(3) **A**   TT 1, 2

If **A** were not recognized as the negation of ¬**A**, this derivation would need the usual additional line for application of the Rule of Double Negation.

(1) ¬**A** → **B**   P
(2) ¬**B**   P
(3) ¬¬**A**   TT 1, 2
(4) **A**   DN 3

We may use these extensions of TT in both the antecedent and consequent, as is illustrated by

(1) ¬**P** → ¬**Q**   P
(2) **Q**   P
(3) **P**   TT 1, 2

An illustration of these ideas in a derivation using mathematical sentences is the following. We want to prove that $x = 0$, and we have three premises.

(1) $x \neq 0$ → $x = y$ P
(2) $x = y$ → $x = z$ P
(3) $x \neq z$ P
(4) $x \neq y$ TT 2, 3
(5) $x = 0$ TT 1, 4

Note that we obtain line (5) from lines (1) and (3) because '$x=0$' is the negation of '$x\neq0$'.

E X E R C I S E 6

A. Using the rule that P is the negation of ¬P, avoid the Rule of Double Negation in the following derivations.

1. Prove: ¬P
   (1) P → ¬Q   P
   (2) Q   P

2. Prove: ¬A
   (1) A → ¬C   P
   (2) B → C   P
   (3) B   P

3. Prove: P
   (1) ¬P → ¬Q   P
   (2) Q   P

4. Prove: A
   (1) ¬A → ¬B   P
   (2) ¬B → ¬C   P
   (3) C   P

5. Prove: ¬S
   (1) P → Q   P
   (2) Q → R   P
   (3) S → ¬R   P
   (4) P   P

6. Prove: ¬A
   (1) A → B   P
   (2) B → C   P
   (3) C → D   P
   (4) ¬D   P

B. Recognizing, for example, that '$x=0$' is the negation of '$x\neq0$', avoid the Rule of Double Negation in the following derivations.

1. Prove: $x=0$
   (1) $x\neq0 \rightarrow x+y\neq y$ P
   (2) $x+y=y$ P

2. Prove: $x\neq0$
   (1) $x=0 \rightarrow x\neq y$ P
   (2) $x=z \rightarrow x=y$ P
   (3) $x=z$ P

3. Prove: $x=y$
   (1) $x\neq y \rightarrow x\neq z$ P
   (2) $x\neq z \rightarrow x\neq0$ P
   (3) $x=0$ P

4. Prove: $x\neq0$
   (1) $x=y \rightarrow x=z$ P
   (2) $x=z \rightarrow x=1$ P
   (3) $x=0 \rightarrow x\neq1$ P
   (4) $x=y$ P

5. Prove: $x\neq y$
   (1) $x=y \rightarrow y=z$ P
   (2) $y=z \rightarrow y=w$ P
   (3) $y=w \rightarrow y=1$ P
   (4) $y\neq1$ P

6. Prove: $x=0$
   (1) $x\neq0 \rightarrow y=1$ P
   (2) $x=y \rightarrow y=w$ P
   (3) $y=w \rightarrow y\neq1$ P
   (4) $x=y$ P

Adjunction and Simplification.   Suppose we are given two sentences as premises. The first is

George is a senior.

The second is

Susie is a junior.

If both sentences are true, then we could put them together in one molecular sentence using the connecting word 'and' and have a true sentence which would read

George is a senior and Susie is a junior.

If both premises are true then the conclusion would have to be true. The rule that permits us to go from the two premises to the conclusion is called the *Rule of Adjunction.* It is abbreviated (A).

In symbolic terms, we can illustrate the rule thus:

From premises	P
	Q
we can conclude	P & Q
or we can conclude	Q & P.

With parentheses the rule looks like this

From premises	(P)
	(Q)
we can conclude	(P) & (Q)
or we can conclude	(Q) & (P).

The parentheses in the conclusion are necessary only if **P** or **Q** are molecular sentences other than negations.

The order that the premises are in does not matter. In our first example we could have concluded 'Susie is a junior and George is a senior'. It would not change the meaning. If we have sentence **Q** as a premise, followed by sentence **P** as a premise, our conclusion can still be **P & Q**, both because the order of lines to which a rule is applied makes no difference and because the order of a conjunction is unimportant.

Below are several examples using the Rule of Adjunction.

a. (1) P	P		*b.* (1) Q & S	P	
(2) ¬R	P		(2) ¬T	P	
(3) P & ¬R	A 1, 2		(3) ¬T & (Q & S)	A 1, 2	

    c. (1) T   P

        (2) U   P

        (3) U & T   A 1, 2

    d. (1) P ∨ Q   P

        (2) Q ∨ R   P

        (3) (P ∨ Q) & (Q ∨ R)   A 1, 2

Now let us look at an example of a rule which is just the reverse of the one we have just learned. We have one premise that says

Mary's birthday is on Friday and mine is on Sunday.

From this premise we can derive two sentences. One conclusion is

Mary's birthday is on Friday.

The other conclusion is

Mine is on Sunday.

If the premise is true, each of these conclusions are always true. The rule that allows us to go from a conjunction to either of the two sentences that are connected by & is called the *Rule of Simplification*. This rule is abbreviated by S.

In symbolic terms the Rule of Simplification is

    From premise   **P & Q**

    we can conclude   **P**

    or we can conclude   **Q**

With parentheses added, the rule is

    From premise   **(P) & (Q)**

    we can conclude   **(P)**

    or we can conclude   **(Q)**.

The parentheses emphasize that the premise *must* be a conjunction. The Rule of Simplification *cannot* be applied to P & Q → R, which means (P & Q) → R; *but it can be applied to* P & (Q → R) to get P or to get Q → R.

Several examples of the use of the Rule of Simplification are

a. (1) (P ∨ Q) & R   P   *b.* (1) Q & S   P

    (2) R   S 1   (2) Q   S 1

c. (1) (P ∨ Q) & R   P   *d.* (1) T & ¬V   P
    (2) P ∨ Q   S 1   (2) ¬V   S 1

                    e. (1) (P & Q) & R   P
                        (2) P & Q   S 1

<h1 style="text-align:center">E X E R C I S E 7</h1>

A.   What conclusion or conclusions can you derive from each of the following sets of premises using rule A or rule S?

 1. A society is a collection of individuals who pursue a way of life, and culture is their way of life.
 2. The atomic number of hydrogen is 1. The atomic number of helium is 2.
 3. Kofi speaks the Twi language. Ama speaks the Ga language.
 4. Tom likes to ski and there is snow in the mountains.
 5. This inference is valid. That is not valid.

 6. (1) Q & R   P   8. (1) R ∨ S   P
    (2)   (2) Q   P

 7. (1) (P ∨ Q) & S   P   9. (1) S   P
    (2)   (2) T   P
                                       (3)

         10. (1) Q & R   P
             (2) S   P
             (3)

B.   Prove that the conclusions below follow logically from the premises given. Show the complete proof.

 1. Prove: ¬S   4. Prove: B & D
    (1) ¬R & T   P   (1) B & C   P
    (2) S → R   P   (2) B → D   P

 2. Prove: A & B   5. Prove: ¬S & Q
    (1) C → A   P   (1) ¬S → Q   P
    (2) C   P   (2) ¬(T & R)   P
    (3) C → B   P   (3) S → T & R   P

 3. Prove: ¬¬Q   6. Prove: A & C
    (1) P & Q   P   (1) A & ¬B   P
                                     (2) ¬C → B   P

Disjunctions as Premises.   Perhaps you have noticed that in the rules that we have learned so far, we have been using conjunctions, conditionals, and negations. The rules have been dealing with the connectives 'and', 'if . . . then . . .', and 'not'. You may also have noticed that one connective and one type of molecular sentence have been left out. The rules have ignored the connective 'or'. We have not used disjunctions in premises to show the use of a rule of inference.

Before a rule is introduced, however, we must consider the *meaning* of a disjunction in logic. In our everyday language there are two possible ways to use the word 'or'. We sometimes mean that either one thing is the case or something else is the case but not both at the same time. This is called the *exclusive* meaning of 'or'. For example, in the sentence

> John either lives north of Valley Road or he lives
> south of Valley Road

we mean that one or the other of the atomic sentences is true and the other one is false.

In logic, however, we will use a broader meaning for the disjunction. This is called the *inclusive* sense. In the inclusive sense, when we use the word 'or', we mean that *at least* one side of the disjunction is the case and perhaps both. Suppose a sign over one gate to a baseball park reads

> Members of the press or photographers enter here.

The meaning of the sentence is the disjunction

> Members of the press enter here or photographers enter here.

It is a disjunction in the inclusive sense, meaning that at least one side is true and both may be. For example, this sentence means that if a person is a member of the press he enters at this gate or if he is a photographer he enters at this gate. Furthermore, the press photographers who are both members of the press and photographers enter at this gate.

In logic, a disjunction means that *at least one* side is true and *perhaps both*. If you remember that we use the word 'or' in the inclusive sense in logic, then you will avoid the kind of mistake which says that if one side of a disjunction is true then the other must be false. Both may be true. The disjunction merely says that *at least one* is true.

With the logical meaning of a disjunction clearly in mind, can you think of a possible rule of inference dealing with a disjunction? Consider the following sentence as a premise.

Either production is increased or the price is increased.

See if you can think of a second premise so that from the two you can derive a valid conclusion. To say the conclusion is *valid* is to say that it follows from the premises by use of a "good" rule of inference. To say that it is a "good" rule is simply to say that whenever the premises are true sentences then the conclusion which follows by that rule is a true sentence. This means valid rules of derivation never allow us to go from true premises to a false conclusion.

Modus Tollendo Ponens.   A rule you might have suggested is the one whose name is *modus tollendo ponens.* Once again, the Latin name tells us something about the rule. It tells us that by *denying* (tollendo) one side of a disjunction we *affirm* (ponens) the other side.

In symbolic form *modus tollendo ponens* may be stated

From the premise	P ∨ Q
and the premise	¬P
we may conclude	Q

or

From the premise	P ∨ Q
and the premise	¬Q
we may conclude	P

The abbreviation for *modus tollendo ponens* is TP.

With the parentheses added *modus tollendo ponens* may be written

From	(P) ∨ Q
and	¬(P)
derive	(Q)

or

From	(P) ∨ (Q)
and	¬(Q)
derive	(P)

Suppose we have as a premise the disjunction

Either this substance includes hydrogen or it includes oxygen.

The second premise says

> This substance does not include hydrogen.

By *modus tollendo ponens* we may then conclude

> This substance includes oxygen.

In order to make clear the *form* of this inference, we can symbolize the previous example. Let

> P = 'This substance includes hydrogen'
> Q = 'This substance includes oxygen'.

The proof of the conclusion is

(1)	P ∨ Q	P
(2)	¬P	P
(3)	Q	TP 1, 2

Note that one premise (the negation) negates or denies one part of the disjunction. The conclusion affirms or states precisely the other part. It does not matter which side of the disjunction, the right or the left, is denied. The disjunction says that at least one side is the case, so if we find that one of the sides is *not* the case, we know that the other must be the case.

A disjunction in logic means that at least one of two sentences is true *and perhaps both*. Suppose you have a premise which tells you that one side of the disjunction is true. Can you conclude anything about the other side? For example, look at the sentence above about hydrogen and oxygen. If the second premise had been 'The substance does include hydrogen', what, if anything, could you conclude about the oxygen in it? You could conclude nothing about it.

Look at the examples below. They are all examples of the use of the rule *modus tollendo ponens*. Notice that our rules are not limited to atomic sentences. Like the other sentence forms, the disjunction may hold between molecular as well as atomic sentences. Notice also that we need parentheses in many of the sentences in order to show which is the major connective.

a.			*b.*		
(1)	Q ∨ R	P	(1)	(P̄ & Q) ∨ S	P
(2)	¬R	P	(2)	¬S	P
(3)	Q	TP 1, 2	(3)	P & Q	TP 1, 2

c. (1) ¬S ∨ T   P   *d.* (1) ¬P ∨ ¬Q   P
   (2) ¬T   P   (2) ¬¬P   P
   (3) ¬S   TP 1, 2   (3) ¬Q   TP 1, 2

       e. (1) (P & Q) ∨ (R & S)   P
         (2) ¬(P & Q)   P
         (3) R & S   TP 1, 2

We also use the fact that **P** is the negation of ¬**P** in applying *modus tollendo ponens*, as is shown in the following examples.

a. (1) Q ∨ ¬R   P   *b.* (1) ¬(P & Q) ∨ S   P
   (2) R   P   (2) P & Q   P
   (3) Q   TP 1, 2   (3) S   TP 1, 2

       c. (1) ¬S ∨ T   P
         (2) S   P
         (3) T   TP 1, 2

EXERCISE 8

A. What conclusion, in the form of an English sentence, can you derive from each of the following sets of premises using rule TP?

1. That man is either a lawyer or he is a politician. He is not a lawyer.
2. The port of New Orleans is either on the Gulf of Mexico or it is on the Atlantic Ocean. It is certainly not on the Atlantic Ocean.
3. Either the internal energy of an atom can be changed continuously or it changes only in steps. The internal energy of an atom cannot be changed continuously.
4. Jim has either finished the book or he is not going to return it to the library today. Jim has not finished the book.
5. Either it is cold and it is rainy or the festival will be held outside. It is not the case that it is both cold and rainy.

B. Derive a conclusion from each of the following sets of premises using *modus tollendo ponens*.

1. (1) ¬Q ∨ R   P   3. (1) ¬T ∨ ¬R   P
   (2) ¬R   P   (2) ¬¬R   P

2. (1) T ∨ (P → Q)   P   4. (1) P ∨ Q   P
   (2) ¬T   P   (2) ¬Q   P

5. (1) (S & T) ∨ R   P
   (2) ¬(S & T)   P

6. (1) (P & Q) ∨ S   P
   (2) ¬S   P

7. (1) ¬Q ∨ R   P
   (2) ¬¬Q   P

8. (1) ¬T   P
   (2) T ∨ ¬S   P

9. (1) ¬(P & Q)   P
   (2) T ∨ (P & Q)   P

10. (1) T ∨ U   P
    (2) ¬T   P

11. (1) S ∨ ¬T   P
    (2) T   P

12. (1) ¬(S & R) ∨ T   P
    (2) S & R   P

13. (1) ¬(P → Q) ∨ R   P
    (2) P → Q   P

C.   Prove that the conclusions follow from the given premises in the exercises below. Show a complete proof.

1. Prove: P
   (1) P ∨ Q   P
   (2) ¬T   P
   (3) Q → T   P

2. Prove: B
   (1) ¬A ∨ B   P
   (2) ¬A → E   P
   (3) ¬E   P

3. Prove: M
   (1) S & P   P
   (2) M ∨ ¬N   P
   (3) S → N   P

4. Prove: A & B
   (1) B   P
   (2) B → ¬D   P
   (3) A ∨ D   P

5. Prove: H
   (1) ¬S   P
   (2) S ∨ (H ∨ G)   P
   (3) ¬G   P

6. Prove: P
   (1) T → P ∨ Q   P
   (2) ¬¬T   P
   (3) ¬Q   P

7. Prove: R
   (1) ¬Q ∨ S   P
   (2) ¬S   P
   (3) ¬(R & S) → Q   P

D.   First symbolize the following premises and conclusions. Then show a proof that the conclusion follows logically from the premises. Remember that when the atomic sentences are already symbolized by mathematical symbols, we need not use the capital letters. Leave the atomic sentences in their mathematical symbols and symbolize the connectives.

1. Either $x=y$ or $x=z$.
   If $x=z$ then $x=6$.
   It is not the case that $x=6$.
   Therefore, $x=y$.
2. Both $1+1=2$ and $2+1=3$.
   Either $3-2=1$ or it is not the case that $2-1=1$.
   If $1+1=2$ then $2-1=1$.
   Therefore, $3-2=1$.
3. If $x \neq 0$ then $x \neq y$.
   Either $x=y$ or $x=z$.
   $x \neq z$.
   Therefore, $x=0$.
4. Either $x=0$ or $x=y$.
   If $x=y$ then $x=z$.
   $x \neq z$.
   Therefore, $x=0$.
5. If $x=y$ then $x=z$.
   If $x=z$ then $x=w$.
   Either $x=y$ or $x=0$.
   If $x=0$ then $x+u=1$.
   $x+u \neq 1$.
   Therefore, $x=w$.

▶ 2.3 *Sentential Derivation*

We have been learning some of the rules of good inference that permit taking a logical step from one set of statements to another statement. From sentence $P \rightarrow Q$ and sentence P, for example, we are permitted to derive the sentence Q.

We have also shown that we can prove that a conclusion does follow logically from a set of premises, even though we cannot go directly from the premises to the conclusion in one step. By going through several steps, each permitted by a rule, we may be able to reach the desired conclusion. If so, we have *proved* that the conclusion follows logically from the given premises.

With a few rules at our command, we began learning the procedure for *formal derivations*. In other words, we have been learning a precise way of proving that arguments are valid. An *argument* is simply a set of sentences as premises and a conclusion derived from these premises.

To say it is *valid* is to say that the conclusion follows logically from the premises. A *formal derivation* is a series of sentences or steps, in which every step is either a premise or is derived directly from the steps that go before it by a definite rule.

In the introduction to this chapter, we compared learning the rules of logic to learning to play a game. Suppose we think of a derivation as a kind of game. We have learned enough rules to do a simple derivation. The derivation or proof is the game and the rules of the game are just the rules of inference. You may make any move, take any step that is permitted by a rule. But you must be able to justify any step taken by stating some definite rule. The goal that we are trying to reach in this game is the stated conclusion. The purpose of each move we make is to get us a step closer to that goal. The starting position from which we begin the game is the set of premises. The premises are justified by the *Rule of Premises* which is

A premise may be introduced at any point in a derivation.

The application of the rules does not depend on using any previous lines.

We have already been using the Rule of Premises to start our derivations. Now that the rule is familiar, the P for Rule of Premises will usually be omitted when a problem is given in symbolized form. In formal derivations, however, you should write a P after each given premise to indicate that the lines are justified by the Rule of Premises.

Summarizing, we begin with a set of premises and the object is to get from these premises to a particular conclusion. Each move we make, each line we write down, must be permitted by a definite rule of inference.

We have learned to do simple derivations. Now let us look at some complicated derivations.

Read the following argument.

Example a.

If the whale is a mammal, then it gets oxygen from the air. If it gets its oxygen from the air, then it does not need gills. The whale is a mammal and its habitat is the ocean. Therefore, it does not need gills.

The conclusion we wish to prove or derive is the sentence 'It does not need gills'. (The word, 'therefore', shows us that the final sentence is the conclusion of the argument.)

The first step in our procedure is to symbolize the argument so that the derivation will be perfectly clear.

Let

W = 'The whale is a mammal'
O = 'It gets its oxygen from the air'
G = 'It does need gills'
H = 'Its habitat is the ocean'.

Then

the first premise is   W → O
the second premise is   O → ¬G
the third premise is   W & H
the conclusion is   ¬G.

The sentential derivation can be written in the way shown below.

(1) W → O   P
(2) O → ¬G   P
(3) W & H   P
(4) W   S 3
(5) O   PP 1, 4
(6) ¬G   PP 2, 5

The first three steps are premises. Steps 4, 5, and 6 are justified by rules of inference applied to previous lines. To the right of each step or line we indicate the way in which we justify that line. For example, since the first three lines are premises, we write the initial letter P to the right of that line. These lines are given and not derived and therefore need no other justification.

Line 4 is derived from line 3 by the rule of simplification. Therefore, we write the abbreviation for the rule (S) to the right of that line, followed by the number of the line from which it was derived. Line 5 is obtained from lines 1 and 4 by *modus ponendo ponens*. If you look at line 1, W → O, and at line 4, W, you can quickly see that *modus ponendo ponens* allows us to obtain O. This move is shown by the abbreviation of the rule's name, PP, and the numbers of the lines from which line 5 was derived. In the same way, line 6 is shown to have been derived by *modus ponendo ponens* from lines 2 and 5.

Since line 6 represents our desired conclusion, the goal of our derivation, we have completed our derivation. We have proved that ¬G follows logically from the three premises of the argument. Thus, since ¬G represents the sentence 'It does not need gills' in the sample argument, we have proved that the conclusion of that argument is valid. This is an example of a *formal derivation*.

In order that every step of the proof is perfectly clear to you and to others who read it, we will keep strictly to the form above while learning to do derivations. One goal in logic is to be precise. To be certain of precision, put down every step you take and why that step is permitted. For each step, first write the number of that line, then the sentence itself, then write what it is that justifies that step, the abbreviation of some rule. If the step is derived from other lines by rule then add the number or numbers of the lines from which it was derived.

Read the following derivation.

Example b.

If the amendment was not approved then the constitution remains as it is. If the constitution remains as it is then we will not add new members to the committee. We will either add new members to the committee or the report will be delayed for a month. But the report will not be delayed for a month. Therefore, the amendment was approved.

Let

A = 'The amendment was approved',
C = 'The constitution remains as it is',
M = 'We will add new members to the committee',
R = 'The report will be delayed for a month'.

Then

(1)	$\neg A \rightarrow C$	P
(2)	$C \rightarrow \neg M$	P
(3)	$M \vee R$	P
(4)	$\neg R$	P
(5)	M	TP 3, 4
(6)	$\neg C$	TT 2, 5
(7)	A	TT 1, 6

The conclusion of the argument is 'The amendment was approved'. Our job, then, is to show that the conclusion follows logically from the four premises of the argument. *First*, we indicate the letter symbols which represent the atomic sentences. *Then*, we symbolize all four premises and show that they are justified in the proof by labeling them with the letter 'P'. The premises are the first four lines in the proof.

Our *next* move is to try to get to the conclusion, which is **A**, by using the rules we have learned. Line 5 is obtained from lines 3 and 4 by *modus tollendo ponens*, TP. Line 6 is derived from lines 2 and 5 by *modus tollendo tollens*, TT. Line 7 is obtained from lines 1 and 6 by *modus tollendo tollens*. We have shown that the conclusion of the argument does follow from the premises by means of a formal derivation.

Read the following derivation.

Example c.

If Tom is seventeen, then Tom is the same age as Jane. If Jim is not as old as Tom, then Jim is not as old as Jane. Tom is seventeen and Jim is as old as Jane. Therefore, Jim is as old as Tom and Tom is the same age as Jane.

Let

$\quad$ **E** = 'Tom is seventeen',
$\quad$ **S** = 'Tom is the same age as Jane',
$\quad$ **T** = 'Jim is as old as Tom',
$\quad$ **J** = 'Jim is as old as Jane'.

Then

(1)	**E** → **S**	P
(2)	¬**T** → ¬**J**	P
(3)	**E** & **J**	P
(4)	**E**	S 3
(5)	**S**	PP 1, 4
(6)	**J**	S 3
(7)	**T**	TT 2, 6
(8)	**T** & **S**	A 5, 7

EXERCISE 9

A. In each of the following examples, prove that the conclusion follows from the premises given. Do each derivation in exactly the way derivations were done in the examples above, with each line numbered, an abbreviation for the rule used, and the numbers of the lines from which each step was derived.

1. If this is a matrilineal society, then the mother's brother is the head of the family. If the mother's brother is the head of the family, then the father does not impose discipline. This is a matrilineal society. Therefore, the father does not impose discipline.

2. Either this rock is an igneous rock or it is a sedimentary rock. This rock is granite. If this rock is granite then it is not a sedimentary rock. Therefore, this rock is an igneous rock.

3. If Jack is taller than Bob, then Mary is shorter than Jean. Mary is not shorter than Jean. If Jack and Bill are the same height, then Jack is taller than Bob. Therefore, Jack and Bill are not the same height.

4. If A did win the race, then either B was second or C was second. If B was second, then A did not win the race. If D was second, then C was not second. A did win the race. Therefore, D was not second.

5. If the clock is fast, then Jones arrived before 10:00 P.M. and he did see Smith's car leave. If Smith is telling the truth, then Jones did not see Smith's car leave. Either Smith is telling the truth or he was in the building at the time of the crime. The clock is fast. Therefore, Smith was in the building at the time of the crime.

B. In the following exercises, the premises are already in symbolic form. Show a complete derivation of the sentence you are asked to prove.

1. Prove: Q
   (1) S → (P ∨ Q)
   (2) S
   (3) ¬P

2. Prove: R
   (1) S → ¬T
   (2) T
   (3) ¬S → R

3. Prove: S & T
   (1) P & R
   (2) P → S
   (3) R → T

4. Prove: ¬S
   (1) T → R
   (2) R → ¬S
   (3) T

5. Prove: T
   (1) P → S
   (2) ¬S
   (3) ¬P → T

6. Prove: S & T
   (1) P → S
   (2) P → T
   (3) P

7. Prove: S
   (1) P ∨ Q
   (2) ¬Q
   (3) P → S

8. Prove: S
   (1) T → R
   (2) ¬R
   (3) T ∨ S

9. Prove: ¬T
   (1) P → S
   (2) P & Q
   (3) (S & R) → ¬T
   (4) Q → R

10. Prove: ¬R
    (1) S ∨ ¬R
    (2) T → ¬S
    (3) T

11. Prove: S
    (1) P → (Q & R)
    (2) P
    (3) T → ¬Q
    (4) T ∨ S

12. Prove: ¬Q
    (1) T ∨ ¬S
    (2) S
    (3) Q → ¬T

13. Prove: Q ∨ R
    (1) S → ¬T
    (2) T
    (3) ¬S → (Q ∨ R)

14. Prove: S
    (1) ¬T ∨ R
    (2) T
    (3) ¬S → ¬R

15. Prove: ¬R
    (1) Q & T
    (2) Q → ¬R
    (3) T → ¬R

C. Give a complete formal proof of the following arguments:

1. Prove: $y+8<12$
   (1) $x+8=12$ ∨ $x\neq4$
   (2) $x=4$ & $y<x$
   (3) $x+8=12$ & $y<x$ → $y+8<12$

2. Prove: $x<4$ & $y<6$*
   (1) $x+2<6$ → $x<4$
   (2) $y<6$ ∨ $x+y\not<10$
   (3) $x+y<10$ & $x+2<6$

---

* For convenience we introduce the notations $\not<$ and $\not>$ for 'is not less than' and 'is not greater than' so that '$\neg(x<y)$' can be written '$x\not<y$' and '$\neg(x>y)$' can be written '$x\not>y$.'

3. Prove: $x=5$ & $x\neq y$
   (1) $x=y \rightarrow x\neq y+3$
   (2) $x=y+3 \lor x+2=y$
   (3) $x+2\neq y$ & $x=5$

4. Prove: $y>z$
   (1) $x=y \rightarrow x=z$
   (2) $x\neq y \rightarrow x<z$
   (3) $x\not< z \lor y>z$
   (4) $y\neq z$ & $x\neq z$

5. Prove: $x<5$
   (1) $x<y \lor x=y$
   (2) $x=y \rightarrow y\neq 5$
   (3) $x<y$ & $y=5 \rightarrow x<5$
   (4) $y=5$

6. Prove: $\tan\theta \neq 0.577$
   (1) $\tan\theta=0.577 \rightarrow \sin\theta=0.500$ & $\cos\theta=0.866$
   (2) $\sin\theta=0.500$ & $\cos\theta=0.866 \rightarrow \cot\theta=1.732$
   (3) $\sec\theta=1.154 \lor \cot\theta\neq 1.732$
   (4) $\sec\theta\neq 1.154$

7. Prove: $\neg(y>7 \lor x=y)$
   (1) $x<6$
   (2) $y>7 \lor x=y \rightarrow \neg(y=4$ & $x<y)$
   (3) $y\neq 4 \rightarrow x\not< 6$
   (4) $x<6 \rightarrow x<y$

8. Prove: $x>6$
   (1) $x>5 \rightarrow x=6 \lor x>6$
   (2) $x\neq 5$ & $x\not< 5 \rightarrow x>5$
   (3) $x<5 \rightarrow x\neq 3+4$
   (4) $x=3+4$ & $x\neq 6$
   (5) $x=3+4 \rightarrow x\neq 5$

9. Prove: $x=4$
   (1) $3x+2y=18$ & $x+4y=16$
   (2) $x=2 \rightarrow {}^-3x+2y\neq 18$
   (3) $x=2 \lor y=3$
   (4) $x\neq 4 \rightarrow y\neq 3$

10. Prove: $x < 3$

   (1) $x + 2 > 5 \quad \rightarrow \quad x = 4$
   (2) $x = 4 \quad \rightarrow \quad x + 4 \nleq 7$
   (3) $x + 4 < 7$
   (4) $x + 2 > 5 \quad \vee \quad (5 - x > 2 \quad \& \quad x < 3)$

▶ *2.4 More About Parentheses*

In the first chapter, we learned that parentheses do the job in logic that several kinds of punctuation and certain words do in our everyday written language. The parentheses show us the grouping of molecular sentences, in which meanings can be quite different if the grouping is different. For example, a sentence symbolized as

$$(A \; \& \; B) \vee C$$

would not have the same meaning as a sentence symbolized as

$$A \; \& \; (B \vee C).$$

In the second grouping, we are certain of **A** and we are certain that either **B** or **C** is the case as well. In the first grouping, we are not certain about any of the sentences. We only know that either **A** & **B** or **C** is the case.

In deriving conclusions from sets of premises, the proper use of parentheses is essential, for otherwise we cannot be certain of the application of our rules. Look at the sentence

$$A \; \& \; Q \vee R.$$

Without parentheses to show grouping, we cannot tell which is the dominant connective nor can we say whether the sentence is a disjunction or a conjunction. We could not know whether we can use the Law of Simplification, or perhaps *modus tollendo ponens*.

We can indicate the major connective by using parentheses. If the sentence is grouped

$$(P \; \& \; Q) \vee R$$

then it is a disjunction and the major connective is 'or'. It is the disjunction whose left member is a molecular sentence (a conjunction) and whose right member is an atomic sentence. If it is grouped

(2) $$P \; \& \; (Q \vee R)$$

then it is a conjunction. We could apply the Rule of Simplification to sentence (2) and derive **P**, but we could *not* derive **P** from sentence (1). Both the meaning of sentences and the proper application of the rules of inference depend upon the correct use of parentheses.

By indicating the grouping of symbolized sentences, the parentheses show us which connective dominates the sentence. You will remember that the conditional connective is stronger than those of conjunction, disjunction, or negation. Where it occurs in a sentence with any of the others, we do not need parentheses to indicate that it is the major connective. We consider 'and' and 'or' to be of equal strength and therefore need parentheses to indicate the dominance of one over the other. Both 'and' and 'or' are stronger than 'not', so that ¬ applies only to the shortest sentence it is placed in front of, *unless* parentheses show that it applies to a longer molecular sentence. It is therefore the dominant connective in

$$\neg(P \rightarrow Q)$$

or in

$$\neg(P \vee Q).$$

EXERCISE 10

A. Does the sentence ¬**Q** & **R** differ in meaning from the sentence ¬(**Q** & **R**)? If so, can you explain the difference in your own words?

B. Suppose our example sentence had been symbolized in the following way: (¬**Q** & **R**)? Does this sentence have the same meaning as either example in exercise **A** above?

C. Suppose we are given as a first premise the sentence **P** → **Q** ∨ **R**. The second premise is the sentence ¬(**Q** ∨ **R**). Can you derive a conclusion from these premises? Could you remove the parentheses from the second premise and derive a conclusion? Explain your answer.

D. For each of the following sentences, tell what the major connective is and what kind of sentence (conjunction, disjunction, negation, or conditional) it is.

1. ¬R ∨ S	5. P ∨ (R & S)
2. P → Q & R	6. ¬(Q & R)
3. (P → Q) & R	7. (P & Q) ∨ (R & S)
4. A & B → C	8. (A → B) & (B → C)

9. ¬A → B ∨ C   12. (P & Q) → (A & B)
10. ¬(P → Q)   13. ¬(P → Q & R)
11. (A & B) ∨ C   14. P → Q ∨ R
                    15. (P → Q) ∨ R

E. Complete the symbolization of the following sentences by adding parentheses where needed to make the symbolized sentence correspond to the name beside it.

1. P → R & S	A conjunction
2. P & R ∨ S	A conjunction
3. A & B → C	A conditional
4. P & R ∨ S	A disjunction
5. ¬P → R	A conditional
6. ¬P → R	A negation
7. ¬P & ¬R	A conjunction
8. ¬P & R	A negation
9. A → B ∨ C	A disjunction
10. A → B ∨ C	A conditional
11. ¬P ∨ Q	A negation
12. ¬P ∨ Q	A disjunction
13. P → Q & R → S	A conjunction
14. ¬¬A → B	A negation
15. ¬P ∨ ¬Q	A disjunction

F. For each of the following sets of premises a conclusion is stated. In some cases the conclusion follows logically only if parentheses are added to show the proper grouping. Add parentheses if needed to make the conclusion valid.

1. P → Q & R   Premise   3. Q & P ∨ S   Premise
   _____   _____
   R   Conclusion   Q   Conclusion

2. P → Q & R   Premise   4. P → Q & S   Premise
   ¬Q & R   Premise   P   Premise
   _____   _____
   ¬P   Conclusion   Q & S   Conclusion

▶ 2.5 *Further Rules of Inference*

At this point our list of rules is rather short and this limits us in the kinds of derivations we can do. With just a few more rules we will be prepared to do far more proofs. You will remember that our rules permit us to go from certain sentences to other sentences which follow from them. Whenever the first sentences are true then the sentences which follow by rules of logic are also true.

Law of Addition.   The *Law of Addition* expresses the fact that if we have a sentence that is true, then the disjunction of that sentence and any other sentence will be true also. If we are given sentence **P**, then sentence **P** ∨ **Q** follows.

To make this seem obvious, think about the meaning of a disjunction. The disjunction **P** ∨ **Q** says that at least one of the sentences connected by the 'or' connective must be true. Remember that *only* one need be true. *Since we were given* **P** *as a true sentence, we know that* **P** ∨ **Q** *must be a true sentence.* This is what we mean, of course, by a logically valid conclusion. When a premise is true, the conclusion which follows from it *must* be true.

Let us look at examples of English sentences just to see how obvious this rule is. If, as a true premise, we are given

This book is blue,

then we know that the following sentence must be true.

Either this book is blue or it is red.

We could also conclude

Either this book is blue or it is old

or

Either this book is blue or it is new

and so on. One part is true in all the above examples and that is all that is needed for a true disjunction.

In symbolic terms, if we have sentence **P**, we can conclude **P** ∨ **Q**, or **P** ∨ **R**, or **S** ∨ **P**, or **T** ∨ **P**, and so forth. The abbreviation for the Law of Addition is LA.

Examples of the law of addition are

a.	(1) **Q**	P
	(2) **Q** ∨ ¬**R**	LA 1
b.	(1) ¬**R**	P
	(2) **S** ∨ ¬**R**	LA 1

$c.$ (1) T & S   P
    (2) (T & S) ∨ R   LA 1

$d.$ (1) T ∨ R   P
    (2) (P & S) ∨ (T ∨ R)   LA 1

Note that the order we use does not matter. From **P** we can derive **P ∨ Q** or we can derive **Q ∨ P**.

EXERCISE 11

A. Give five sentences that result from the following premise

Some games are easy to learn.

B. If the conclusions follow from the premises in the examples below, write the word 'valid'. If it is valid, complete the proof by telling the rule or rules used and the lines to which a rule was applied. If it does not apply a rule of inference you have learned, mark an 'X' through the conclusion.

1. (1) P   P
   Therefore: P ∨ Q

2. (1) Q   P
   Therefore: Q ∨ ¬R

3. (1) P   P
   (2) P ∨ Q → R   P
   Therefore: R

4. (1) Q ∨ R → S   P
   (2) R   P
   Therefore: S

5. (1) T   P
   (2) S ∨ T → Q
   Therefore: Q

6. (1) ¬R   P
   (2) ¬S ∨ ¬R → ¬P   P
   Therefore: ¬P

7. (1) ¬T   P
   Therefore: ¬T ∨ ¬P

8. (1) ¬P   P
   Therefore: Q ∨ ¬P

9. (1) P   P
   Therefore: P & ¬Q

10. (1) R & S → T   P
    (2) R
    Therefore: T

C. Show a derivation of the following conclusions from the sets of given premises. Do a formal proof, showing the number of each step, the justification for every line by the abbreviation for the rule used, and the numbers of the lines from which you derived each step.

1. Prove: $T \lor S$
   (1) $Q \lor T$
   (2) $Q \to R$
   (3) $\neg R$

2. Prove: $R \lor \neg T$
   (1) P
   (2) $\neg R \to \neg P$

3. Prove: $R \lor \neg S$
   (1) $S \,\&\, Q$
   (2) $T \to \neg Q$
   (3) $\neg T \to R$

4. Prove: Q
   (1) $\neg S$
   (2) $T \to S$
   (3) $\neg T \lor R \to Q$

5. Prove: U
   (1) $P \,\&\, \neg T$
   (2) $S \to T$
   (3) $S \lor Q$
   (4) $Q \lor P \to U$

6. Prove: $T \lor Q$
   (1) $S \to P \,\&\, Q$
   (2) S
   (3) $P \,\&\, Q \to T$

D.   Give a formal proof of the following arguments.

1. Prove: $y \not< 4 \lor x > 2$
   (1) $x > 3 \lor y \not< 4$
   (2) $x > 3 \to x > y$
   (3) $x \not> y$

2. Prove: $x > y \lor y \not< 6$
   (1) $x > y \lor x > 5$
   (2) $x \not> 5 \lor y \not< 6$
   (3) $x + y = 1 \,\&\, x > y$

3. Prove: $x \neq 3 \lor x > 2$
   (1) $x + 2 \neq 5 \lor 2x = 6$
   (2) $x + 2 \neq 5 \to x \neq 3$
   (3) $2x - 2 = 8 \to 2x \neq 6$
   (4) $x + 3 = 8 \,\&\, 2x - 2 = 8$

4. Prove: $\tan 30° = 0.577 \lor \cos 60° = 0.5$
   (1) $\sin 30° = 0.5 \to \csc 30° = 2.0$
   (2) $\sin 30° = 0.5$
   (3) $\csc 30° = 2.0 \to \tan 30° = 0.577$

5. Prove: $x = 5 \,\&\, x \neq 4$
   (1) $x = 2 \to x < 3$
   (2) $x \neq 4 \,\&\, x \not< 3$
   (3) $x \neq 2 \lor x > 4 \to x = 5$

6. Prove: $x=2$
   (1) $Dx^3=3x^2$ & $D3=0$
   (2) $Dx^3=3x^2$ $\rightarrow$ $Dx^2=2x$
   (3) $Dx^2=2x$ $\vee$ $Dx^3=12$ $\rightarrow$ $x=2$

7. Prove: $x=3$
   (1) $x-2=1$ & $2-x\neq1$
   (2) $x=1$ $\rightarrow$ $2-x=1$
   (3) $x=1$ $\vee$ $x+2=5$
   (4) $x+2=5$ $\vee$ $x-2=1$ $\rightarrow$ $x=3$

8. Prove: $y=x$ $\vee$ $y>x$
   (1) $y<6$ $\rightarrow$ $y<x$
   (2) $y\not<6$ $\vee$ $x=5$ $\rightarrow$ $y>x$
   (3) $y\not<x$

9. Prove: $y<3$ $\vee$ $x>5$
   (1) $y<4$ & $x=y+3$
   (2) $\neg(x\neq y+3)$ $\rightarrow$ $x>2$
   (3) $y\not>2$ $\rightarrow$ $x\not>2$
   (4) $y>2$ $\vee$ $y=3$ $\rightarrow$ $x>5$

10. Prove: $(x=4$ $\vee$ $y\neq8)$ & $x<3$
    (1) $x=y$ $\vee$ $x<y$
    (2) $y=x+4$
    (3) $(x<3$ $\vee$ $x>5)$ & $y=x+4$ $\rightarrow$ $y\neq8$
    (4) $x\neq y$
    (5) $y=6$ $\vee$ $x<y$ $\rightarrow$ $x<3$

Law of Hypothetical Syllogism.   First let us look at an example of the *Law of the Hypothetical Syllogism*, which we abbreviate HS. From the premises

   (1) If the day is warm, then Judy goes swimming.
   (2) If Judy goes swimming, then she does her homework after dinner.

we can conclude

   (3) If the day is warm, then she does her homework after dinner.

To symbolize the argument, let

> D = 'The day is warm'
> S = 'Judy goes swimming'
> H = 'She does her homework after dinner'.

Then

(1)	$D \rightarrow S$	P
(2)	$S \rightarrow H$	P
(3)	$D \rightarrow H$	HS

The conclusion is a conditional sentence. Both premises are conditional sentences. The conclusion does not tell us that the day is warm, nor that Judy does her homework after dinner. It only tells what will happen *if* the day is warm. Look at the two premises and imagine that the antecedent of the first premise is true. If so, then the consequent of the second would surely follow. That is exactly what our conditional conclusion is saying. In terms of symbols, if we have the sentence $D \rightarrow S$ and the sentence $S \rightarrow H$ and if we also had the sentence D then we could apply *modus ponendo ponens* twice and get H. In short if D then H, or $D \rightarrow H$. But this is just what we concluded from $D \rightarrow S$ and $S \rightarrow H$ by HS.

In symbolic form the Law of the Hypothetical Syllogism is

from	$P \rightarrow Q$
and	$Q \rightarrow R$
we can conclude	$P \rightarrow R.$

You may find it convenient to think of three steps in applying HS. First, as a rough check, notice that it requires two conditionals. Second, as a careful check, be sure that the antecedent of one of the conditionals is precisely the same as the consequent of the other. Third, the conclusion is then another conditional whose antecedent is the other antecedent of one of the premises and whose consequent is the consequent of the other premise.

In the examples of the Law of the Hypothetical Syllogism given below note that some of the antecedents and consequents are molecular sentences. The form, however, is the same.

a.	(1)	$\neg P \rightarrow \neg Q$	P
	(2)	$\neg Q \rightarrow \neg R$	P
	(3)	$\neg P \rightarrow \neg R$	HS 1, 2

    b. (1) ¬P → Q ∨ R   P
       (2) Q ∨ R → ¬T   P
       (3) ¬P → ¬T   HS 1, 2

    c. (1) S → T   P
       (2) T → R ∨ Q   P
       (3) S → R ∨ Q   HS 1, 2

    d. (1) (P → Q) → R   P
       (2) R → (Q & T)   P
       (3) (P → Q) → (Q & T)   HS 1, 2

EXERCISE 12

A. What, if anything, can you conclude from the following sets of sentences by the Law of Hypothetical Syllogism?

1. If water freezes, then its molecules form crystals. If the molecules form crystals then water expands.
2. If Tom drives at the rate of 50 miles per hour, then in 9 hours he will travel 450 miles.
   If in 9 hours he travels 450 miles, then he will travel 90 miles farther than in the same period yesterday.
3. If Mr. Lincoln is elected, then the Southern States will surely secede. If the Southern States secede, then a civil war will result.
4. If a beam of photons penetrates gas in a cloud chamber, then the photons eject electrons from the atoms of gas. If photons eject electrons from atoms of gas, then the energy of the light passes into the kinetic energy of the electrons.
5. If representation in the Senate is according to population, then New York has more Senators than Nevada. If New York has more Senators than Nevada, then New York has more than two Senators.

B. Translate the arguments in Section **A** above into logical symbols and prove that your conclusions follow logically from the premises.

C. Example 5 in Section **A** above shows the *hypothetical* character of the premises of a hypothetical syllogism. The premises are all true in that example. But what about the factual truth of the atomic sentences in Example 5? Can you add one more premise that you know to be a true sentence (*in fact*) to the argument in Example 5 in order to prove that its atomic sentences are not true sentences. Show how you could then prove the negation of those atomic sentences by means of a formal symbolic proof.

D. Use the Law of Hypothetical Syllogism (HS) with a formal proof to reach a conclusion from the following sets of premises.

1. (1) Q → ¬P
   (2) ¬P → R

2. (1) P → R & ¬S
   (2) R & ¬S → T

3. (1) S ∨ T → R ∨ Q
   (2) R ∨ Q → ¬P

4. (1) S → ¬T
   (2) ¬T → ¬R

E. Show a formal derivation of the following conclusions from the given premises.

1. Prove: ¬T
   (1) (Q → R) & P
   (2) R → T
   (3) (Q → R) → ¬T

2. Prove: P
   (1) ¬R
   (2) ¬P → Q
   (3) Q → R

3. Prove: Q
   (1) ¬R → S
   (2) S → P & Q
   (3) R → T
   (4) ¬T

F. Give formal proofs of the following arguments.

1. Prove: $(2+2)+2=6 \rightarrow 3+3=6$
   (1) $(2+2)+2=6 \rightarrow 3\times2=6$
   (2) $3\times2=6 \rightarrow 3+3=6$

2. Prove: $5x-4=3x+4 \rightarrow x=4$
   (1) $5x-4=3x+4 \rightarrow 5x=3x+8$
   (2) $2x=8 \rightarrow x=4$
   (3) $5x=3x+8 \rightarrow 2x=8$

3. Prove: $z>6 \lor z<y$
   (1) $x>y \rightarrow x>z$
   (2) $\neg(z>6) \rightarrow \neg(x>y) \rightarrow z<7)$
   (3) $x>z \rightarrow z<7$

4. Prove: $x=6 \quad \vee \quad x>6$
   (1) $x \neq y \quad \rightarrow \quad y<x$
   (2) $(x>5 \quad \rightarrow \quad y<x) \quad \rightarrow \quad y=5$
   (3) $y \neq 5 \quad \vee \quad x=6$
   (4) $x>5 \quad \rightarrow \quad x \neq y$

5. Prove: $x>y$
   (1) $x \neq y \quad \rightarrow \quad x>y \quad \vee \quad x<y$
   (2) $x>y \quad \vee \quad x<y \quad \rightarrow \quad x \neq 4$
   (3) $x<y \quad \rightarrow \quad \neg(x \neq y \quad \rightarrow \quad x \neq 4)$
   (4) $x \neq y$

6. Prove: $(y \neq 0 \quad \vee \quad x<z) \quad \& \quad (x<y \quad \rightarrow \quad x=0)$
   (1) $x<y \quad \rightarrow \quad x=0$
   (2) $y=0 \quad \rightarrow \quad x \not< y$
   (3) $x<y \quad \& \quad z=3$
   (4) $x<y \quad \rightarrow \quad x<z$

7. Prove: $\neg(z \neq 5) \quad \vee \quad z>5$
   (1) $x=3 \quad \rightarrow \quad x>y$
   (2) $x \neq 3 \quad \rightarrow \quad z=5$
   (3) $(x=3 \quad \rightarrow \quad x<z) \quad \rightarrow \quad x \not< z$
   (4) $x>y \quad \rightarrow \quad x<z$

8. Prove: $x \neq 3 \quad \vee \quad 4>x$
   (1) $5x=20 \quad \rightarrow \quad x=4$
   (2) $2x=6 \quad \vee \quad x \neq 3$
   (3) $2x=6 \quad \rightarrow \quad \neg(5x-3=17 \quad \rightarrow \quad x=4)$
   (4) $5x-3=17 \quad \rightarrow \quad 5x=20$

9. Prove: $y+z=8$
   (1) $z=5 \quad \rightarrow \quad ((y=3 \quad \rightarrow \quad y+z=8) \quad \& \quad z>y)$
   (2) $(xy+z=11 \quad \rightarrow \quad x=2) \quad \rightarrow \quad (y=3 \quad \& \quad z=5)$
   (3) $xy=6 \quad \rightarrow \quad x=2$
   (4) $xy+z=11 \quad \rightarrow \quad xy=6$

10. Prove: $x+z=3 \quad \rightarrow \quad y=3$
    (1) $(x+y=5 \quad \rightarrow \quad y=3) \quad \vee \quad x+z=3$
    (2) $z \neq 1 \quad \vee \quad (x+z=3 \quad \rightarrow \quad x+y=5)$
    (3) $x+y \neq 5 \quad \& \quad z=1$

Law of Disjunctive Syllogism.   The *Law of Disjunctive Syllogism*, abbreviated DS, begins with a disjunction and two conditionals. Look at the example,

> Either it is raining or the field is dry.
> If it is raining, then we will play indoors.
> If the field is dry, then we will play baseball.

What conclusion could you draw from these sentences? The conclusion is that we will either play indoors or we will play baseball. The conclusion is another disjunction.

Now let us symbolize the argument above to get a clear picture of the form of a disjunctive syllogism. We let

> R = 'It is raining'
> D = 'The field is dry'
> P = 'We will play indoors'
> B = 'We will play baseball'.

This argument is symbolized

(1)	$R \lor D$	P
(2)	$R \rightarrow P$	P
(3)	$D \rightarrow B$	P
(4)	$P \lor B$	DS

In symbols the Law of Disjunctive Syllogism can be stated

from	$P \lor Q$
and	$P \rightarrow R$
and	$Q \rightarrow S$
derive	$R \lor S$
or derive	$S \lor R$.

You may find it convenient to think of three steps in applying DS. First, as a rough check, notice that it requires two conditionals and a disjunction. Second, as a careful check, be sure that the two antecedents of the two conditionals are precisely the same as the two members of the disjunction. Third, the conclusion is another disjunction whose members are precisely the same as the two consequents of the two conditionals.

Several examples of the Law of the Disjunctive Syllogism follow. In the conclusion, either member of the disjunction may be stated first.

1. (1) ¬P ∨ Q   P
   (2) ¬P → ¬R   P
   (3) Q → S   P
   (4) ¬R ∨ S   DS 1, 2, 3

2. (1) P ∨ Q   P
   (2) P → ¬R   P
   (3) Q → ¬S   P
   (4) ¬S ∨ ¬R   DS 1, 2, 3

3. (1) ¬P ∨ ¬Q   P
   (2) ¬P → R   P
   (3) ¬Q → S   P
   (4) R ∨ S   DS 1, 2, 3

4. (1) P ∨ ¬Q   P
   (2) P → ¬R   P
   (3) ¬Q → S   P
   (4) ¬R ∨ S   DS 1, 2, 3

5. (1) P ∨ Q   P
   (2) P → R   P
   (3) Q → ¬S   P
   (4) ¬S ∨ R   DS 1, 2, 3

EXERCISE 13

A. What can you conclude from each of the following sets of premises by the Law of Disjunctive Syllogism? State your conclusion in an English sentence.

1. Either Bob has the majority or Dick has the majority. If Bob has the majority, then Bob will be treasurer. If Dick has the majority, then Dick will be treasurer.

2. This number is either a positive number or it is a negative number. If it is a positive number, then it is greater than zero. If it is a negative number, then it is less than zero.

3. This rock is either limestone or it is granite. If it is limestone, then it is sedimentary. If it is granite, then it is igneous.

4. The camera was either acquired legally by the seller or the camera is stolen merchandise. If the camera was acquired legally by the seller, then it is my camera. If the camera is stolen merchandise, then Tom is its legal owner.

5. Either the plant is a green plant or it is a nongreen plant. If it is a green plant, then it makes its own food. If it is a nongreen plant, then it depends upon materials from other plants for its food.

B. Symbolize the arguments in Section **A** and prove that your conclusions follow logically from the premises.

C. Use the Law of Disjunctive Syllogism (DS) to reach a conclusion from each of the following sets of premises.

1. (1) P ∨ ¬Q
   (2) ¬Q → R
   (3) P → ¬S

2. (1) Q ∨ R
   (2) Q → ¬S
   (3) R → ¬T

3. (1) ¬T ∨ ¬S
   (2) ¬S → P
   (3) ¬T → Q

4. (1) (R & S) ∨ T
   (2) (R & S) → ¬Q
   (3) T → P

D. Show a full formal derivation of the following conclusions from the premises given.

1. Prove: R & (P ∨ Q)
   (1) P ∨ Q
   (2) Q → R
   (3) P → T
   (4) ¬T

2. Prove: T
   (1) P ∨ ¬R
   (2) ¬R → S
   (3) P → T
   (4) ¬S

3. Prove: ¬Q & S
   (1) S & ¬R
   (2) R ∨ ¬T
   (3) Q → T

4. Prove: S
   (1) P → Q
   (2) Q → ¬R
   (3) R
   (4) P ∨ (T & S)

5. Prove: ¬T & ¬P
   (1) ¬S ∨ ¬R
   (2) ¬R → ¬T
   (3) ¬S → P
   (4) ¬P

E. Give a formal proof of each of the following arguments:

1. Prove: $x=3$ ∨ $x=2$
   (1) $x+y=7$ → $x=2$
   (2) $y-x=2$ → $x=3$
   (3) $x+y=7$ ∨ $y-x=2$

2. Prove: $x > 2 \quad \lor \quad x = 2$
   (1) $x < y \quad \rightarrow \quad x = 2$
   (2) $x < y \quad \lor \quad x \not< y$
   (3) $x \not< y \quad \rightarrow \quad x > 2$

3. Prove: $y = 1$
   (1) $2x + y = 7 \quad \rightarrow \quad 2x = 4$
   (2) $2x + y = 5 \quad \rightarrow \quad y = 1$
   (3) $2x + y = 7 \quad \lor \quad 2x + y = 5$
   (4) $2x \neq 4$

4. Prove: $y = 1 \quad \lor \quad y = 9$
   (1) $\neg(x = 2 \quad \lor \quad x = 8) \quad \rightarrow \quad x = 6$
   (2) $2x + 3y = 21 \quad \& \quad x \neq 6$
   (3) $x = 2 \quad \rightarrow \quad y = 9$
   (4) $x = 8 \quad \rightarrow \quad y = 1$

5. Prove: $\neg(x \not< z) \quad \lor \quad \neg(z \neq 6)$
   (1) $x > 5 \quad \lor \quad y \not< 6$
   (2) $y \not< 6 \quad \rightarrow \quad x < z$
   (3) $x > 5 \quad \rightarrow \quad y < z$
   (4) $y \not< z \quad \& \quad z = 6$

6. Prove: $x \neq 4 \quad \lor \quad x > y$
   (1) $y = 0 \quad \rightarrow \quad xy = 0$
   (2) $y = 0 \quad \lor \quad y \not< 1$
   (3) $xy = 0 \quad \lor \quad xy > 3 \quad \rightarrow \quad x \neq 4$
   (4) $y \not< 1 \quad \rightarrow \quad xy > 3$

7. Prove: $y < 12 \quad \lor \quad x < 0$
   (1) $x < y \quad \lor \quad y < x$
   (2) $y < x \quad \rightarrow \quad x > 6$
   (3) $x < y \quad \rightarrow \quad x < 7$
   (4) $(x > 6 \quad \lor \quad x < 7) \quad \rightarrow \quad y \not> 11$
   (5) $y > 11 \quad \lor \quad x < 0$

8. Prove: $x^2 = 4 \quad \lor \quad x^2 = 9$
   (1) $2x^2 - 10x + 12 = 0 \quad \& \quad x < 4$
   (2) $x^2 - 5x + 6 = 0 \quad \rightarrow \quad x = 2 \quad \lor \quad x = 3$
   (3) $x = 2 \quad \rightarrow \quad x^2 = 4$
   (4) $x = 3 \quad \rightarrow \quad x^2 = 9$
   (5) $2x^2 - 10x + 12 = 0 \quad \rightarrow \quad x^2 - 5x + 6 = 0$

9. Prove: $x+1 \not> y \quad \vee \quad x \not> 4$
   (1) $(y=5 \quad \rightarrow \quad x<y) \quad \& \quad x>1$
   (2) $y>5 \quad \vee \quad y=5$
   (3) $x<y \quad \vee \quad y>4 \quad \rightarrow \quad x+1 \not> y \quad \& \quad y<9$
   (4) $y>5 \quad \rightarrow \quad y>4$

10. Prove: $x=4$
    (1) $x=5 \quad \vee \quad x<y$
    (2) $x>3 \quad \vee \quad z<2 \quad \rightarrow \quad z<x \quad \vee \quad y=1$
    (3) $x<y \quad \rightarrow \quad z<2$
    (4) $x=5 \quad \rightarrow \quad x>3$
    (5) $z<x \quad \rightarrow \quad x=4$
    (6) $y=1 \quad \rightarrow \quad \neg(x>3 \quad \vee \quad z<2)$

Law of Disjunctive Simplification. If someone says, 'The Giants will win or the Giants will win', we can conclude that he believes simply, 'The Giants will win'. In symbolized form the argument is

$$G \vee G$$

Therefore,

$$G.$$

This is an example of the Law of Disjunctive Simplification, which we abbreviate DP.

Some examples of this law are

a. (1) P $\vee$ P	P	*b.* (1) $\neg$Q $\vee$ $\neg$Q	P
(2) P	DP 1	(2) $\neg$Q	DP 1

c. (1) (P & Q) $\vee$ (P & Q)   P
   (2) P & Q   DP 1

Notice that the possibilities for simplifying a disjunction are much more limited than those for simplifying a conjunction. In the case of a disjunction the two sentences must be exactly the same.

An important application of the Law of Disjunctive Simplification occurs when a disjunctive syllogism has the following special form,

$$P \vee Q$$
$$P \rightarrow R$$
$$Q \rightarrow R$$

Therefore,

R ∨ R.

In this special case we can simplify the conclusion **R** ∨ **R** to **R**. For if
R ∨ **R** is true, then it must be the case that **R** is true. The inference from
R ∨ **R** *to* **R** is an example of the Law of Disjunctive Simplification.

E X E R C I S E 14

A. Use the Laws of Disjunctive Syllogism and Disjunctive Simplification to reach a conclusion from each of the following sets of symbolized premises.

1. (1) S ∨ ¬T
   (2) ¬T → R
   (3) S → R

2. (1) ¬Q → ¬S
   (2) P ∨ ¬Q
   (3) P → ¬S

3. (1) S → ¬R
   (2) T → ¬R
   (3) S ∨ T

4. (1) ¬R → S
   (2) Q ∨ ¬R
   (3) Q → S

B. If you think the following argument "makes sense", can you
explain why you think so?

> If Bob is elected, then Jim will be appointed chairman of the
> committee.
> If Tom is elected, then Jim will be appointed chairman of the
> committee.
> Either Bob will be elected or Tom will be elected.
> Therefore, Jim will be appointed chairman of the committee.

C. What conclusions can you draw from the following sets of premises
using the Laws of Disjunctive Syllogism and Disjunctive Simplification?

1. There are either three members of the committee or there are
   five members.
   If there are three members, then there will not be a tie vote.
   If there are five members, then there will not be a tie vote.
2. If this closed figure has three sides, then it is a triangle.
   If this closed figure has three angles, then it is a triangle.
   This closed figure either has three sides or it has three angles.
3. Either the Bears or the Tigers will finish first.
   If the Bears finish first, then the Knights will be third.
   If the Tigers finish first, then the Knights will be third.

D. Show a formal proof of the following conclusions from the sets of premises given.

1. Prove: $\neg T$ & S   2. Prove: Q   3. Prove: $\neg S$ & R
   (1) $P \rightarrow \neg Q$ (1) $Q \lor S$ (1) $S \rightarrow P$
   (2) $P \lor R$ (2) $S \rightarrow T$ (2) $\neg P$ & $\neg T$
   (3) $R \rightarrow \neg Q$ (3) $\neg T$ (3) $\neg T \rightarrow R$
   (4) $T \rightarrow Q$
   (5) S

E. Symbolize the following argument and then prove that the conclusion can be derived logically from the premises.

> Jane either scored 65 on the test or she scored 70.
> If Jane scored 65 on the test, then she did not get B.
> If she scored 70, then she did not get B.
> If Jane studies, then she gets B on her tests.
> Therefore, Jane did not study.

F. Give formal proof of each of the following arguments:

1. Prove: $x < 4$
   (1) $x = y \lor x > y$
   (2) $x < 4 \lor x \nless z$
   (3) $x = y \rightarrow x < z$
   (4) $x > y \rightarrow x < z$

2. Prove: $x = 1$
   (1) $2x + y = 5 \rightarrow 2x = 2$
   (2) $2x + y = 5 \lor y = 3$
   (3) $2x = 2 \rightarrow x = 1$
   (4) $y = 3 \rightarrow 2x = 2$

3. Prove: $x = 2$
   (1) $x < 3 \lor x > 4$
   (2) $x < 3 \rightarrow x \neq y$
   (3) $x > 4 \rightarrow x \neq y$
   (4) $x < y \lor x \neq y$
      $\rightarrow x \neq 4$ & $x = 2$

4. Prove: $x^2 = 9$
   (1) $x = (+3) \rightarrow 2x^2 = 18$
   (2) $x = (+3) \lor x = (-3)$
   (3) $x = (-3) \rightarrow 2x^2 = 18$
   (4) $2x^2 = 18 \rightarrow x^2 = 9$

5. Prove: $\neg(x \neq 5)$
   (1) $z > x \rightarrow x \neq 7$
   (2) $x < 6 \lor x = 3$
   (3) $x = 3 \rightarrow z > x$
   (4) $x < 6 \rightarrow z > x$
   (5) $x = 7 \lor x = 5$

6. Prove: $y = z \lor x \neq 5$
   (1) $x = y \rightarrow x < z$
   (2) $x = 5 \rightarrow x \nless z$
   (3) $x = 3 \rightarrow x < z$
   (4) $x \nless y \rightarrow x = y$
   (5) $x = 3 \lor x \nless y$

7. Prove: $x^2=9 \quad \vee \quad x^2>9$
    (1) $x=3 \quad \vee \quad x=4$
    (2) $x=3 \quad \rightarrow \quad x^2-7x+12=0$
    (3) $x=4 \quad \rightarrow \quad x^2-7x+12=0$
    (4) $x^2-7x+12=0 \quad \rightarrow \quad x>2$
    (5) $x^2<9 \quad \rightarrow \quad x>2$
    (6) $x^2 \nless 9 \quad \rightarrow \quad x^2=9 \quad \vee \quad x^2>9$

8. Prove: $\cos\theta = \frac{1}{2}\sqrt{3} \quad \vee \quad \csc\theta = 2$
    (1) $\theta = 150° \quad \rightarrow \quad \sin\theta = \frac{1}{2}$
    (2) $\theta = 30° \quad \vee \quad \theta = 150°$
    (3) $\sin\theta = \frac{1}{2} \quad \rightarrow \quad \csc\theta = 2$
    (4) $\theta = 30° \quad \rightarrow \quad \sin\theta = \frac{1}{2}$

Commutative Laws. These rules will probably seem especially obvious to you, yet we cannot take any step for granted. We must have an explicit rule for each move. The following argument is an example of the use of one of the Commutative Laws.

> Galileo died in 1642 and Isaac Newton was born in 1642. Therefore, Isaac Newton was born in 1642 and Galileo died in 1642.

In symbolic terms,

from   **P & Q**

derive   **Q & P.**

It seems very obvious because we know that the order of the atomic sentences in a conjunction does not affect the meaning of that molecular sentence. No doubt everyone would agree that whenever **P & Q** is true, **Q & P** is also true. You remember that this is precisely our standard for a good rule of inference: that whenever the premises are true the conclusion is true.

Another type of molecular sentence to which the Commutative Laws apply is the following:

> Either x is greater than five or x is equal to five.

Is the following conclusion valid?

> Either x is equal to five or x is greater than five.

The answer is "yes" because it must be true when the premise is true. Let us symbolize this argument to illustrate the form of the Commutative Laws here. Let

P = 'x is greater than five'

Q = 'x is equal to five'.

The argument is

$$\begin{array}{ll} \text{from} & \text{P} \vee \text{Q} \\ \text{derive} & \text{Q} \vee \text{P}. \end{array}$$

The Commutative Laws apply to conjunctions and disjunctions. This means that a change in the order of the two members of the conjunction or the disjunction does not alter its meaning. The Commutative Laws do not apply to conditionals. Further examples are given below.

a. (1) P & ¬Q   P   *c.* (1) ¬P ∨ ¬Q   P
     (2) ¬Q & P   CL 1   (2) ¬Q ∨ ¬P   CL 1

b. (1) ¬P & Q   P
     (2) Q & ¬P   CL 1

You may abbreviate these rules as

CL.

EXERCISE 15

A.   Apply the Commutative Laws to the following sentences to obtain a different sentence.

1. Addition is a binary operation and multiplication is a binary operation.
2. Either an unbalanced force acts on a body or the velocity of the body is not changed.
3. Jim's uncle is either a senator or he is a representative in the State Legislature.
4. Tony lives on Maple Street and Cathy lives on Sixth Street.
5. Either hydrogen is a liquid or it is a gas.

B.   What can you conclude from the following premises by using the Commutative Laws?

1. (1) ¬P ∨ Q   P   4. (1) ¬T ∨ ¬S   P
2. (1) ¬R & ¬S   P   5. (1) Q ∨ R   P
3. (1) P & S   P

C.   Prove that the following conclusions are valid by showing a complete formal derivation.

1. Prove: S & Q
  (1) P ∨ T
  (2) ¬T
  (3) P → Q & S

2. Prove: ¬(R & ¬T)
  (1) (R & ¬T) → ¬S
  (2) P → S
  (3) P & Q

3. Prove: S & R
  (1) (R & S) ∨ P
  (2) Q → ¬P
  (3) T → ¬P
  (4) Q ∨ T

4. Prove: R ∨ Q
  (1) S → R
  (2) S ∨ T
  (3) ¬T

5. Prove: T
  (1) P → Q
  (2) Q → R
  (3) (P → R) → ¬S
  (4) S ∨ T

6. Prove: ¬S
  (1) ¬T ∨ ¬S
  (2) ¬Q → T
  (3) Q → ¬R
  (4) R

7. Prove: R
  (1) S → R ∨ T
  (2) ¬¬S
  (3) ¬T

8. Prove: ¬Q & P
  (1) T → P & ¬Q
  (2) T ∨ ¬R
  (3) R

9. Prove: T
  (1) P ∨ ¬R
  (2) ¬S
  (3) P → S
  (4) ¬R → T

10. Prove: ¬P
  (1) R → T
  (2) S → Q
  (3) T ∨ Q → ¬P
  (4) R ∨ S

11. Prove: $y \not< 4$ & $x < y$
  (1) $x > y$ ∨ $x < 4$
  (2) $x < 4$ → $x < y$ & $y \not< 4$
  (3) $x > y$ → $x = 4$
  (4) $x \neq 4$

12. Prove: $y > 3$ & $y < 5$
  (1) $x = 3$ ∨ $y = 3$
  (2) $x > 2$ ∨ $x + y \not> 5$
  (3) $y = 3$ ∨ $x = 3$ → $x + y > 5$
  (4) ¬($y < 5$ & $y > 3$) → $x \not> 2$

13. Prove: $x < 3$ & $y = 7$
   (1) $x < 3$ & $y > 6$
   (2) $y \neq 7$ $\rightarrow$ $\neg(x = 2$ & $y > x)$
   (3) $y > 6$ & $x < 3$ $\rightarrow$ $y > x$ & $x = 2$

14. Prove: $x = 1$ & $(y < 1$ $\vee$ $y < 2)$
   (1) $x + 2y = 5$ $\vee$ $3x + 4y = 11$
   (2) $x \not< 2$ $\vee$ $x > y$ $\rightarrow$ $y < 2$ $\vee$ $y < 1$
   (3) $3x + 4y = 11$ $\rightarrow$ $x = 1$
   (4) $x > y$ $\vee$ $x \not< 2$
   (5) $x + 2y = 5$ $\rightarrow$ $x = 1$

15. Prove: $x < 6$
   (1) $x > y$ $\vee$ $x < 6$
   (2) $x > y$ $\rightarrow$ $x > 4$
   (3) $x > 4$ $\rightarrow$ $x = 5$ & $x < 7$
   (4) $x < 6$ $\rightarrow$ $x = 5$ & $x < 7$
   (5) $x < 7$ & $x = 5$ $\rightarrow$ $z > x$ $\vee$ $y < z$
   (6) $x > y$ $\rightarrow$ $\neg(y < z$ $\vee$ $z > x)$

De Morgan's Laws.   In English there are often several sentences that express the same meaning. For example,

(1) It is not raining and it is not sunny

can also be expressed as

(2) It is not either raining or sunny.

The logical form of (2) is clearer if it is written

(2′) Not either it is raining or it is sunny.

If (1) and (2) mean the same thing in English then in logic it would be valid to conclude (2) from (1) or to conclude (1) from (2). In symbols this would be

(*a*) from $\neg P$ & $\neg Q$ we can conclude $\neg(P \vee Q)$

and

(*b*) from $\neg(P \vee Q)$ we can conclude $\neg P$ & $\neg Q$.

So (*b*) says that if we do not have either **P** or **Q**, then we do not have **P** and we do not have **Q**. The rule that permits this conclusion is one of the rules known as *De Morgan's Laws.*

The premises of (*a*) and (*b*) are two of the molecular sentential forms to which De Morgan's Laws can apply. Their application to two other forms can be seen if we consider the two equivalent sentences,

> (3) Either it is not hot or it is not snowing.

and

> (4) It is not both hot and snowing.

The logical form of (4) is clearer when written as

> (4′) Not both it is hot and it is snowing.

Since (3) and (4) mean the same, either can be inferred from the other. Therefore, in logical symbols

> (*c*) from ¬P ∨ ¬Q we can conclude ¬(P & Q),
> (*d*) from ¬(P & Q) we can conclude ¬P ∨ ¬Q.

Thus (*c*) and (*d*) are two more examples of the application of De Morgan's Laws. Another is

> (*e*) from P & Q we can conclude ¬(¬P ∨ ¬Q).

And yet another is

> (*f*) from ¬(P ∨ ¬Q) we can conclude ¬P & ¬¬Q
> or ¬P & Q.

Very frequently there is more than one possible form for the conclusion. Fortunately, by studying the different sentential forms to which De Morgan's Laws apply we will find a pattern that is always followed. It will then be possible to state De Morgan's Laws in one rule of operation that works on any form of premise that could be used and gives whatever form of the conclusion we want. We shall now carefully examine the six sentential forms shown to find this pattern.

a. $\dfrac{\text{¬P \& ¬Q}}{\text{¬(P ∨ Q)}}$ *d.* $\dfrac{\text{¬(P \& Q)}}{\text{¬P ∨ ¬Q}}$

b. $\dfrac{\text{¬(P ∨ Q)}}{\text{¬P \& ¬Q}}$ *e.* $\dfrac{\text{P \& Q}}{\text{¬(¬P ∨ ¬Q)}}$

c. $\dfrac{\text{¬P ∨ ¬Q}}{\text{¬(P \& Q)}}$ *f.* $\dfrac{\text{¬(P ∨ ¬Q)}}{\text{¬P \& Q}}$

First, notice that the premise is always one of the following three forms: (1) a conjunction as in (*a*) and (*e*); (2) a disjunction as in (*c*); or

(3) a negation as in (*b*), (*d*), and (*f*). When it is a negation, it must be the negation of a conjunction as in (*d*) or the negation of a disjunction as in (*b*) and (*f*). The premise is never a conditional, nor the negation of a conditional, nor the negation of a negation.

Second, notice that when the premise is a negation the conclusion is not a negation as in (*b*), (*d*), (*f*). And when the premise is not a negation then the conclusion is a negation as in (*a*), (*c*), (*e*). The whole formula is negated either by adding a negation symbol in front of the formula or by taking the negation symbol away from in front of the formula.

Third, notice that & always changes to ∨ and ∨ changes to &.

Fourth, notice that each of the members of the conjunction or disjunction always either gains or loses a negation symbol. Thus each member is negated in applying De Morgan's Laws.

All that we need notice to state De Morgan's Laws, DL, *as a rule of operation* are the following steps:

1. Change & to ∨ or ∨ to &;
2. Negate each member of the disjunction or conjunction;
3. Negate the whole formula.

In the next example these three steps have been performed each time De Morgan's Laws are applied; but by making different choices as to how to do each of the three negatings, conclusions in different forms are obtained. Notice that there is a choice as to how to negate ¬(¬P & Q) or how to negate ¬P, but no choice as to how to negate **Q**.

(1)	¬(¬P & **Q**)	P
(2)	P ∨ ¬**Q**	DL 1
(3)	¬¬P ∨ ¬**Q**	DL 1
(4)	¬¬(P ∨ ¬**Q**)	DL 1
(5)	¬¬(¬¬P ∨ ¬**Q**)	DL 1

Some other examples of the application of De Morgan's Laws are

$$a. \quad \frac{\neg(P \ \& \ \neg Q)}{\neg P \ \lor \ \neg\neg Q} \qquad\qquad c. \quad \frac{\neg\neg P \ \lor \ \neg Q}{\neg(\neg P \ \& \ Q)}$$

$$b. \quad \frac{\neg(\neg P \ \& \ \neg Q)}{\neg\neg P \ \lor \ \neg\neg Q} \qquad\qquad d. \quad \frac{\neg(P \ \lor \ \neg Q)}{\neg P \ \& \ \neg\neg Q}$$

$$e. \quad \frac{\neg\neg P \ \& \ \neg Q}{\neg(\neg P \ \lor \ Q)}$$

EXERCISE 16

A. What can you conclude from the following premises using De Morgan's Laws?

1. Either arachnids are not insects or they do not have eight legs.
2. It is not the case that either air is a good conductor of heat or that water is a good conductor of heat.
3. It is not the case that a number is both greater than zero and that it is negative.
4. The Mississippi River does not flow northward and the Nile River does not flow southward.
5. It is not the case that either bats are birds or that porpoises are fish.

B. Translate the premises and conclusions of Section **A** above into logical symbols and prove that your conclusion follows logically from the premise.

C. Apply De Morgan's Laws to the following sentences to derive conclusions.

1. $\neg(P \,\&\, Q)$
2. $\neg R \lor \neg T$
3. $\neg(\neg R \,\&\, S)$
4. $\neg G \lor \neg H$
5. $\neg\neg S \lor \neg T$
6. $\neg P \,\&\, \neg Q$
7. $\neg\neg P \,\&\, \neg Q$
8. $\neg(C \lor D)$
9. $\neg(\neg P \,\&\, \neg Q)$
10. $\neg S \lor \neg\neg T$
11. $R \,\&\, S$
12. $\neg(\neg A \,\&\, \neg B)$

D. Show a complete formal proof for each of the following symbolized arguments.

1. Prove: $\neg S$
   - (1) $\neg(P \,\&\, Q)$
   - (2) $\neg Q \to T$
   - (3) $\neg P \to T$
   - (4) $S \to \neg T$

2. Prove: $\neg(A \lor B)$
   - (1) $C \,\&\, \neg D$
   - (2) $C \to \neg A$
   - (3) $D \lor \neg B$

3. Prove: $R \,\&\, Q$
   - (1) $\neg S \to \neg(P \lor \neg T)$
   - (2) $T \to Q \,\&\, R$
   - (3) $\neg S$

4. Prove: $\neg R$
   - (1) $P \to \neg Q$
   - (2) $\neg Q \to \neg S$
   - (3) $(P \to \neg S) \to \neg T$
   - (4) $R \to T$

5. Prove: **D**
   (1) ¬A → B
   (2) C → B
   (3) C ∨ ¬A
   (4) ¬B ∨ D

6. Prove: ¬T
   (1) T → P & S
   (2) Q → ¬P
   (3) R → ¬S
   (4) R ∨ Q

7. Prove: ¬P
   (1) R → ¬P
   (2) (R & S) ∨ T
   (3) T → (Q ∨ U)
   (4) ¬Q & ¬U

8. Prove: **R & Q**
   (1) P ∨ Q
   (2) S → Q & R
   (3) P → S
   (4) Q → S

9. Prove: **G ∨ ¬H**
   (1) E ∨ F → ¬H
   (2) J → E
   (3) K → F
   (4) J ∨ K

10. Prove: **S & T**
    (1) ¬(P ∨ ¬R)
    (2) Q ∨ P
    (3) R → S
    (4) (Q & S) → (T & S)

E.   Give a complete formal proof for each of the following arguments.

1. Prove: $\neg(x=3 \quad \& \quad x<2)$
   (1) $\neg(x<2 \quad \& \quad x=3)$

2. Prove: $\neg(x=5 \quad \& \quad y=4)$
   (1) $y \neq 3$
   (2) $x+y=8 \quad \rightarrow \quad y=3$
   (3) $x+y=8 \quad \lor \quad x \neq 5$

3. Prove: $x>y$
   (1) $\neg(y>5 \quad \& \quad x \neq 6)$
   (2) $x=6 \quad \rightarrow \quad x>y$
   (3) $y \not> 5 \quad \rightarrow \quad x>y$

4. Prove: $y=2 \quad \& \quad x>y$
   (1) $x \not< y$
   (2) $x<y \quad \lor \quad \neg(x \not> 3 \quad \lor \quad x+y<5)$
   (3) $x>3 \quad \rightarrow \quad \neg(x \not> y \quad \lor \quad y \neq 2)$

5. Prove: $\neg(x=2 \quad \lor \quad y<5)$
   (1) $\neg(y-x=2 \quad \lor \quad x+y \not> 8)$
   (2) $\neg(x>y \quad \lor \quad y<5)$
   (3) $x=2 \quad \rightarrow \quad x+y \not> 8$

6. Prove: $x < y \quad \vee \quad y \neq 4$
   (1) $x = 1 \quad \rightarrow \quad x < y$
   (2) $x^2 - 4x + 3 = 0 \quad \rightarrow \quad x = 1 \quad \vee \quad x = 3$
   (3) $\neg(x = y \quad \vee \quad x^2 - 4x + 3 \neq 0)$
   (4) $x = 3 \quad \rightarrow \quad x < y$

7. Prove: $2 + 3 \neq 3 \times 3 \quad \vee \quad 2 \times 3 \neq 1 \times 4$
   (1) $2 \times 3 = 1 \times 4 \quad \& \quad 2 + 3 = 3 \times 3 \quad \rightarrow \quad 2 + 3 = 6$
   (2) $2 + 3 \neq 6 \quad \vee \quad 2 \times 3 = 5$
   (3) $2 \times 3 \neq 5$

8. Prove: $x - y \neq 2$
   (1) $\neg(x > y \quad \& \quad x + y > 7)$
   (2) $x \not> y \quad \rightarrow \quad x < 4$
   (3) $x + y \not> 7 \quad \rightarrow \quad x < 4$
   (4) $x - y = 2 \quad \rightarrow \quad x \not< 4$

9. Prove: $x = 1$
   (1) $\neg(z < 3 \quad \vee \quad x > y) \quad \& \quad y = 2$
   (2) $x \not< y \quad \vee \quad x = 1$
   (3) $x > z \quad \rightarrow \quad x > y$
   (4) $x \not> z \quad \rightarrow \quad x < y$

10. Prove: $\neg(x = y \quad \vee \quad y \not> 1)$
   (1) $y \neq 1 \quad \& \quad y \not< 1$
   (2) $y \not> 1 \quad \rightarrow \quad y < 1 \quad \vee \quad y = 1$
   (3) $x = 3 \quad \vee \quad x > 3$
   (4) $x > 3 \quad \rightarrow \quad x \neq y$
   (5) $x = 3 \quad \rightarrow \quad x \neq y$

▶ 2.6 *Biconditional Sentences*

Thus far we have been analyzing molecular sentences by using only
four sentential connectives. There is another sentential connective that
we shall be using later on. This connective is 'if and only if'. Sentences
that use this connective are called *biconditional sentences*. The symbol
we shall use for this connective is

$$\leftrightarrow.$$

The symbol itself tells us a good deal about the biconditional
sentence. The sign looks like two conditional signs going in opposite

directions. Actually a biconditional sentence is very much like two conditional sentences. To illustrate this let us look at an example in our everyday language:

> These fields are flooded if and only if the water
> reaches this height.

In symbolized form the sentence would be

$$P \leftrightarrow Q,$$

where **P** is the name of the first sentence and **Q** is the name of the last sentence. We can read this sentence: **P** if and only if **Q**.

The biconditional sentence $P \leftrightarrow Q$ has the same force as two conditional sentences, first, $P \rightarrow Q$ and second, $Q \rightarrow P$. The English sentence means that if the water reaches a certain height then the fields are flooded. It also means that if the fields are flooded then the water has reached a certain height.

Thus we have a new rule that permits us to derive both $P \rightarrow Q$ and $Q \rightarrow P$ from $P \leftrightarrow Q$. It also lets us derive $P \leftrightarrow Q$ from both $P \rightarrow Q$ and $Q \rightarrow P$. We will call this rule the *Law for Biconditional Sentences* (LB). In symbols it allows the following arguments.

a. $P \leftrightarrow Q$	P		*c.* $P \leftrightarrow Q$		P
$\overline{P \rightarrow Q}$	LB		$\overline{(P \rightarrow Q) \mathbin{\&} (Q \rightarrow P)}$		LB
b. $P \leftrightarrow Q$	P		*d.* $P \rightarrow Q$		P
$\overline{Q \rightarrow P}$	LB		$Q \rightarrow P$		P
			$\overline{P \leftrightarrow Q}$		LB

We shall adopt the rule that the biconditional 'if and only if' is stronger than any of the other connectives. Thus, without parentheses, we know that

$$P \rightarrow Q \leftrightarrow S \mathbin{\&} P$$

is a biconditional rather than a conditional or conjunction. To make it a conditional, parentheses are necessary, as shown in

$$P \rightarrow (Q \leftrightarrow S \mathbin{\&} P).$$

The consequent of this conditional is a biconditional. If we want the consequent to be a conjunction, additional parentheses are needed as in

$$P \rightarrow ((Q \leftrightarrow S) \mathbin{\&} P).$$

Since the ↔ dominates the other connectives unless parentheses show otherwise, the following formulas are also biconditionals:

$a.$ ¬P ↔ ¬Q $b.$ P & Q ↔ S $c.$ S ∨ T ↔ ¬P
   R & S ↔ P & Q   P ∨ Q ↔ R ∨ S   ¬T ↔ S

EXERCISE 17

A. Name each connective you find in the following sentences:

1. This material will melt if and only if it is subjected to very intense heat.
2. We will skate if and only if the ice is not too thin.
3. Tim will walk if and only if the car is gone.
4. Sound travels if and only if there is a transmitting medium.
5. This figure has four interior angles if and only if it has four sides.

B. Symbolize completely the sentences in Section **A** above, telling what atomic sentence is symbolized by each of your letter symbols.

C. Symbolize completely the premises and conclusion of each of the following arguments and give a formal derivation:

1. This piece of legislation will be passed this session if and only if it is supported by the majority party. Either it is supported by the majority party or the governor opposes it. If the governor opposes it, then it will be delayed in committee deliberations. Therefore, either this piece of legislation will be passed this session or it will be delayed in committee deliberation.
2. The sun rises and sets if and only if the earth rotates. The earth rotates and the moon revolves about the earth. Therefore, the sun rises and sets or the climate is very hot or cold.
3. $3 \times 5 = 12$ ↔ $5 + 5 + 5 = 12$
   $4 \times 4 \neq 13$
   $5 + 5 + 5 = 12$ → $4 \times 4 = 13$
   Therefore, $3 \times 5 \neq 12$
4. The land can be cultivated if and only if a means of irrigation is supplied. If the land can be cultivated, then it will be worth three times its present value.
   Therefore, if a means of irrigation is supplied, then the land will be worth three times its present value.

5. A liquid is an acid if and only if it turns blue litmus paper red. A liquid turns blue litmus paper red if and only if it contains free hydrogen ions.

Therefore, a liquid is an acid if and only if it contains free hydrogen ions.

6. If it is not the case that if an object floats on water then it is less dense than water, then you can walk on water.

But you cannot walk on water.

If an object is less dense than water, then it can displace its own weight in water.

If it can displace its own weight in water, then the object will float on water.

Therefore, an object will float on water if and only if it is less dense than water.

D. Give a complete formal proof of each of the following arguments.

1. Prove: $2 \times 5 = 5 + 5 \ \rightarrow \ 2 \times 4 = 4 + 4$
   (1) $2 \times 4 = 4 + 4 \ \leftrightarrow \ 2 \times 5 = 5 + 5$

2. Prove: $x = 4 \ \leftrightarrow \ 3x + 2 = 14$
   (1) $3x + 2 = 14 \ \leftrightarrow \ 3x = 12$
   (2) $3x = 12 \ \leftrightarrow \ x = 4$

3. Prove: $x + y = 5$
   (1) $3x + y = 11 \ \leftrightarrow \ 3x = 9$
   (2) $3x = 9 \ \rightarrow \ 3x + y = 11 \ \leftrightarrow \ y = 2$
   (3) $y \neq 2 \ \lor \ x + y = 5$

4. Prove: $\neg(2x \neq 8 \ \ \& \ \ x \neq 3)$
   (1) $2x = 6 \ \leftrightarrow \ x = 3$
   (2) $2x = 8 \ \leftrightarrow \ x = 4$
   (3) $2x = 6 \ \lor \ x = 4$

5. Prove: $\neg(y = 2 \ \ \& \ \ x + 2y \neq 7)$
   (1) $5x = 15 \ \leftrightarrow \ x = 3$
   (2) $5x = 15 \ \ \& \ \ 4x = 12$
   (3) $x = 3 \ \rightarrow \ x + 2y = 7$

6. Prove: $x \nless y \ \ \& \ \ x \neq y$
   (1) $y \not> x \ \leftrightarrow \ x = y \ \lor \ x < y$
   (2) $\neg(y < 1 \ \lor \ y \not> x)$

7. Prove: $x<y \quad \leftrightarrow \quad y>x$
   (1) $y>x \quad \leftrightarrow \quad x<y$

8. Prove: $x<y \quad \& \quad y=6$
   (1) $x<y \quad \leftrightarrow \quad y>4$
   (2) $y=6 \quad \leftrightarrow \quad x+y=10$
   (3) $y>4 \quad \& \quad \neg(x+y\neq 10)$

9. Prove: $xy\neq 0$
   (1) $y>x \quad \leftrightarrow \quad x=0$
   (2) $xy=0 \quad \leftrightarrow \quad x=0$
   (3) $y\not> x$

10. Prove: $\neg(x<y \quad \& \quad x=1)$
    (1) $x=y \quad \rightarrow \quad x\not< y$
    (2) $y=0 \quad \leftrightarrow \quad x\not< y$
    (3) $x=0 \quad \vee \quad xy=0 \quad \rightarrow \quad y=0$
    (4) $(x=y \quad \rightarrow \quad y=0) \quad \rightarrow \quad x=0$

▶ 2.7 *Summary of Rules of Inference*

We have learned that one important business of logic is that of inference or the derivation of conclusions from sets of premises. In order to perform deductions we learned that we need to make use of certain rules of inference. These rules operate like the rules of any game. They permit us to make certain moves. Every move permitted by the rules is a step in inference, a sentence that can be derived if we are given certain other sentences. At this point in our study of logic we have learned fourteen different rules of inference, enough so that we are able to make long and rather complicated derivations. In formal proofs or derivations we justify every step of inference taken by referring to the particular rule of inference which permits that step. We indicate this rule by putting the abbreviation for its name to the right of the step of inference. It is also necessary to indicate the numbers of the lines in the inference from which each step was derived.

The rules of logic are not, of course, just any arbitrarily chosen rules. They are such that they allow us to make only valid inferences. A valid inference is one that follows logically from the premises. This means that when the premises are true the conclusion that follows must also be true. The point of the rules of inference is to insure that when we are given sets of true sentences the conclusions that can be derived from those sentences will also be true.

To go further in the study of logic it is essential that you become very familiar not only with the idea of valid inference itself but with each particular rule of inference which permits a logical step to be taken. If you do not know these logical rules well, you will be unable to plan a strategy that will help to lead you to the conclusion desired.

Immediately following this section is a table in which is given the name of each rule and the form of that rule. It may be used as a reference table. Remember that the form of the inference is the same whether the parts of the molecular sentence themselves are simple atomic sentences or whether they are themselves molecular sentences.

In the table, we will avoid filling out the form of a formal proof and simply return to using a long line. We put under the line the sentence that results from applying the rule to the sentence or sentences above the line. The sentences above the line may, in a proof, be either premises or derived lines.

Table of Rules of Inference

Modus ponendo ponens (PP)

$$P \rightarrow Q$$
$$P$$
$$\overline{}$$
$$Q$$

Modus tollendo tollens (TT)

$$P \rightarrow Q$$
$$\neg Q$$
$$\overline{}$$
$$\neg P$$

Modus tollendo ponens (TP)

$P \lor Q$	$P \lor Q$
$\neg P$	$\neg Q$
Q	P

Double Negation (DN)

P	$\neg\neg P$
$\neg\neg P$	P

Rule of Simplification (S)

$P \& Q$	$P \& Q$
P	Q

Rule of Adjunction (A)

P	P
Q	Q
$P \& Q$	$Q \& P$

Law of Hypothetical Syllogism (HS)

$$P \rightarrow Q$$
$$Q \rightarrow R$$
$$\overline{}$$
$$P \rightarrow R$$

Law of Addition (LA)

P	Q
$P \lor Q$	$P \lor Q$

De Morgan's Laws (DL)
1. Change & to $\lor$ or $\lor$ to &
2. Negate each member of the conjunction or disjunction
3. Negate the whole formula.

Law of Disjunctive Simplification (DP)

$$\frac{P \lor P}{P}$$

Law of Disjunctive Syllogism (DS)

$$\frac{\begin{array}{c} P \lor Q \\ P \to R \\ Q \to S \end{array}}{R \lor S}$$ or $$\frac{\begin{array}{c} P \lor Q \\ P \to R \\ Q \to S \end{array}}{S \lor R}$$

Commutative Laws (CL)

$$\frac{P \,\&\, Q}{Q \,\&\, P} \qquad \frac{P \lor Q}{Q \lor P}$$

Law for Biconditional Sentences (LB)

$$\frac{P \leftrightarrow Q}{P \to Q} \qquad \frac{P \leftrightarrow Q}{Q \to P}$$

$$\frac{\begin{array}{c} P \to Q \\ Q \to P \end{array}}{P \leftrightarrow Q} \qquad \frac{P \leftrightarrow Q}{(P \to Q) \,\&\, (Q \to P)}$$

Rule of Premises (P)
A premise may be introduced at any point in a derivation.

CHAPTER THREE
TRUTH AND VALIDITY

▶ 3.1 *Introduction*

In our study of logic, we have been concerned with proving the validity of conclusions given certain premises. We learned that if premises are true statements then the conclusions that follow logically from them must be true. We can find individual arguments expressed in English sentences in which we know the premises to be true statements and we know the conclusions to be true. Does this mean that the form of inference that goes from premises of these forms to a conclusion of this form is logically valid? The answer is *no*. There may be other cases of arguments of the same form in which premises are true but the conclusion is false. A single case in which true premises lead to a true conclusion is not sufficient to prove that a given inference is valid.

We have been able to claim that an argument is valid only if we could support our claim by referring to a particular rule of inference in the case of each derived sentence. In Chapter Two we learned a number of rules of inference that allow us to make this claim of validity. The question arises: Is it possible that there are valid sentential inferences that our rules would not be sufficient to support?

Suppose that someone suggests as a rule of inference that if we have sentence $P \rightarrow Q$, then we can derive sentence $\neg P \lor Q$. In other words, he is saying that if $P \rightarrow Q$ is a true sentence, then the sentence $\neg P \lor Q$ will always be true. The inference is, in fact, a valid inference. But if we refer to the list of rules of inference that have been learned so far we will not find one that allows us to go immediately from this premise to that conclusion.

There are many such cases of valid inference for which we have not introduced a particular named rule. Since any suggested inference is either valid or invalid we shall want to be able to prove either validity or invalidity beyond doubt. If we can derive the conclusion by use of our rules, then it is valid. If we can find one case of this form in which the premises are true and the suggested conclusion is false, then

110

we know that it is invalid because true premises lead only to true valid conclusions.

But suppose after trying a long time we have failed to find a proof. *That* does not prove it is invalid. And suppose after trying a long time we have failed to find some case showing that it must be invalid. *That* does not prove it valid. What is needed in these cases is a general method for proving validity or invalidity. The purpose of this chapter and the next is to introduce a method that will be adequate for dealing with any possible example of sentential inference.

▶ 3.2 *Truth Value and Truth-Functional Connectives*

We shall begin with the idea that every sentence must have a *truth value;* every sentence must be either true or false. The truth value of a true sentence is *true* and the truth value of a false sentence is *false*. Every sentence, atomic or molecular, has one of these two possible truth values.

It happens that if we know the truth values of the atomic sentences within molecular sentences then we are able to tell the truth values of the molecular sentences themselves. This is because the five connectives we have been using to form molecular sentences are *truth-functional* connectives. This means that the truth or falsity of a molecular sentence depends completely upon the truth or falsity of the atomic sentences that make it up. To determine the truth or falsity of any molecular sentence we need know only the truth or falsity of its atomic sentences and which connectives combine them. We shall take each sentential connective separately and see how this works.

Conjunction. 'And' is a truth-functional connective, so that you can tell the truth value of sentence **P** & **Q** if you know the truth values of sentence **P** and sentence **Q**. The conjunction of two sentences is true if and only if both sentences are true. Therefore, if **P** & **Q** is to be a true sentence then **P** must be true, and **Q** must be true. This is the case no matter what two sentences we join by means of the connective 'and'. In logic we may combine any two sentences to form a conjunction. We do not require that the subject matter of one be in any way related to the subject matter of the other.

There are four possible combinations of truth value for sentences **P** and **Q**. Remembering that the truth of the conjunction **P** & **Q** depends upon those truth values, let us find the combinations that will make the conjunction **P** & **Q** a true sentence. The four possible combinations are

> **P** is true and **Q** is true.
> **P** is true and **Q** is false.
> **P** is false and **Q** is true.
> **P** is false and **Q** is false.

The rule of usage for conjunctions is: *The conjunction of two sentences is true if and only if both sentences are true.* Thus we conclude

> If **P** is true and **Q** is true, then **P** & **Q** is true.
> If **P** is true and **Q** is false, then **P** & **Q** is false.
> If **P** is false and **Q** is true, then **P** & **Q** is false.
> If **P** is false and **Q** is false, then **P** & **Q** is false.

EXERCISE 1

A.   Jack says, "My birthday is in August and Jane's is in the following month". We find that his birthday is in August but that he was mistaken about Jane's birthday because it is in November. Is Jack's sentence a true or false one? Can you explain your answer in terms of the rules of usage for conjunction?

B.   Tell whether **P** & **Q** is true (T) or false (F) in each of the following cases:

1. Where **P** is a true sentence and **Q** is a true sentence.
2. Where **P** is a true sentence but **Q** is a false sentence.
3. Where both **P** and **Q** are false sentences.
4. Where neither **P** nor **Q** are false sentences.
5. Where **P** is a false sentence but **Q** is a true sentence.

Negation.   The connective 'not' is truth-functional because the truth or falsity of a negation depends entirely upon the truth or falsity of the sentence it negates. *The negation of a true sentence is false and the negation of a false sentence is true.*

Let us apply what has just been said to an example of a negation in ordinary language. Consider the negation,

> John is not Mike's brother.

To know the truth or falsity of that sentence we need only to know the truth or falsity of the sentence,

John is Mike's brother.

If the second sentence is a true one, then the first sentence, its negation, must be false. If the second sentence is a false sentence, then the first sentence must be true.

We are concerned with the negation ¬P. Sentence P may be true or it may be false. The possible truth values for the negation ¬P are

If P is true, then ¬P is false.
If P is false, then ¬P is true.

EXERCISE 2

A. Tell whether P & ¬Q is true (T) or false (F) in each of the following cases:

1. Where P is false and Q is true.
2. Where P is true and Q is false.
3. Where both P and Q are true.
4. Where both P and Q are false.
5. Where P is true and ¬Q is true.

Disjunction. The connective 'or' is also a truth-functional connective. But in considering the truth or falsity of any disjunction we should keep in mind that we are using the *inclusive* sense of the word 'or'. This means that in any disjunction at least one of the two sentences is true and perhaps both. The requirement is that *at least* one side be true. The rule of usage is: *The disjunction of two sentences is true if and only if at least one of the sentences is true.* Again this means that to know the truth or falsity of sentence P ∨ Q we must know about the truth or falsity of sentences P and Q.

Consider the English sentence,

Either Andy won a letter in track or he won a letter in football.

To know whether that sentence is true or false we need to know whether the sentences 'Andy won a letter in track' and 'He won a letter in football' are true or false sentences. If at least one of them is a true sentence, then the entire disjunction is true. Furthermore, if both of the sentences are true, then the disjunction is still a true sentence. If the sentences are both false, then of course the disjunction must be false.

Since the disjunction controls two sentences, again we have four possible combinations of truth and falsity. For the disjunction P ∨ Q the four possibilities are, as in the case of conjunction,

P is true and Q is true.
P is true and Q is false.
P is false and Q is true.
P is false and Q is false.

In determining the truth value of P ∨ Q, we find

If P is true and Q is true, then P ∨ Q is true.
If P is true and Q is false, then P ∨ Q is true.
If P is false and Q is true, then P ∨ Q is true.
If P is false and Q is false, then P ∨ Q is false.

EXERCISE 3

A.   Tell whether P ∨ Q is true (T) or false (F) in each of the following cases:

1. Where P is false but Q is true.
2. Where P and Q are both true.
3. Where P is true and Q is false.
4. Where P is false and Q is false.
5. Where both P and Q are false sentences.

B.   Tell whether ¬R ∨ ¬S is true (T) or false (F) in each of the following cases:

1. Where R is true and S is true.
2. Where R is false and S is true.
3. Where R is true and S is false.
4. Where R is false and S is false.
5. Where both R and S are true.

Conditional Sentences.   If we know the truth or falsity of P and Q then we also know the truth or falsity of P → Q. This is because the truth or falsity of **P** → **Q** is a function of or depends upon the truth or falsity of the antecedent and of the consequent. 'If . . . then . . .' is a truth-functional connective.

This is saying that we know whether a sentence such as

(1) If there is an eclipse then the stars come out

is true or false if we know whether the sentences 'There is an eclipse' and 'The stars come out' are true or false.

Again there are four possible combinations of the truth or falsity of the two atomic sentences. Only two of the possibilities arise very often in the ordinary use of language. First, when 'There is an eclipse' is true and 'The stars come out' is true (1) would be true. Second, when the antecedent 'There is an eclipse' is true but the consequent 'The stars come out' is false then (1) is false. But suppose the antecedent is false, there is no eclipse. In logic, since this is no test of (1) no matter whether the consequent, 'The stars come out' is true or false, we call the conditional (1) true. This fits its use in science and mathematics too. Thus the conditional (1) is a scientific generalization that only says what will happen *if* there is an eclipse. It cannot be disproved (shown to be false) if there is no eclipse. So in logic if the antecedent of a conditional is false then the whole conditional is considered true, no matter whether the consequent is true or false.

Examining the four cases above we see that the whole conditional also turns out true whenever the consequent 'The stars come out' is true no matter whether the antecedent is true or false. This is because the only case when the whole conditional is false is when the antecedent 'There is an eclipse' is true while the consequent 'The stars come out' is false.

If the rule just described seems odd to you it is because we usually think that in a conditional sentence the factual truth of the consequent depends in some way upon the factual truth of the antecedent. In logic this is not the case. The subject matter of the antecedent need not be related in any way to the subject matter of the consequent. We can consider the truth value of

(2) If the day is cold, then $3+3=6$

even though the two atomic sentences seem to have nothing to do with one another. Since the consequent of (2) is obviously a true sentence, (2) itself must be true. The rule of usage is: *A conditional sentence is false if the antecedent is true and the consequent is false; otherwise the conditional sentence is true.* As in the case of the other molecular sentences that contain both sentence P and sentence Q, P → Q has four possibilities of truth and falsity. They are

> P is true and Q is true.
> P is true and Q is false.
> P is false and Q is true.
> P is false and Q is false.

Since the truth value of P → Q is determined solely by the truth or falsity of sentence P and sentence Q, we can analyze its truth value in the following way:

> If P is true and Q is true, then P → Q is true.
> If P is true and Q is false, then P → Q is false.
> If P is false and Q is true, then P → Q is true.
> If P is false and Q is false, then P → Q is true.

By looking over the list above you will notice that whenever the antecedent of a conditional sentence is false, then the conditional sentence is a true one. Notice also that whenever the consequent of a conditional sentence is true then the conditional sentence is a true sentence. The only case in which a conditional sentence is false is the case in which its antecedent is true but its consequent is false.

EXERCISE 4

A.   Mary says, "If this paper is correct then $5 \times 2 = 10$." We know that $5 \times 2 = 10$. For the purposes of logic, what can you say about the truth value of Mary's statement? Why?

B.   Tell whether P → Q is true (T) or false (F) in each of the following cases:

1. Where P and Q are both false.
2. Where P is true and Q is false.
3. Where P is true and Q is true.
4. Where P is false and Q is true.
5. Where P and Q are both true.

C.   Tell whether R → ¬S is true (T) or false (F) in each of the following cases:

1. Where R and S are both false.
2. Where R and S are both true.

3. Where **R** is true and **S** is false.
4. Where **R** is false and **S** is true.
5. Where neither **R** nor **S** is a true sentence.

Equivalence: Biconditional Sentences.   You have been introduced to molecular sentences that contain the connective 'if and only if'. These
sentences, the biconditionals, are also called equivalences. An example
of an equivalence is

> (1) You are permitted to vote if and only if you are
>   registered.

Since 'if and only if' is a truth-functional connective, the truth or
falsity of the equivalence depends upon the truth or falsity of its parts.
This means that the truth value of (1) depends upon the truth or
falsity of 'You are permitted to vote' and 'You are registered'.

If both sentences 'You are permitted to vote' and 'You are
registered' are true sentences then the biconditional sentence (1) is
true. Furthermore, if both sentences or members of the biconditional
are false then sentence (1) is a true sentence. On the other hand, if one
of the members of a biconditional sentence is false while the other
member is true then the biconditional sentence is a false sentence. The
rule of usage should be clear to you if you remember that a biconditional sentence or equivalence $P \leftrightarrow Q$ has essentially the same meaning
as two conditionals, $P \rightarrow Q$ and $Q \rightarrow P$. This means that whenever
you have a biconditional sentence with one true member and one
false member then one of the implications that it contains will have a
true antecedent and a false consequent. The entire sentence, therefore,
must be a false one.

The rule of usage for equivalences is

> *A biconditional sentence is true if and only if its two
> members are either both true or both false.*

Again, since the biconditional sentence contains two members both of
which may be either true or false, there are four possible combinations
of truth or falsity. They are

> **P** is true and **Q** is true.
> **P** is true and **Q** is false.
> **P** is false and **Q** is true.
> **P** is false and **Q** is false.

The truth values for any biconditional sentence **P** ↔ **Q** may be determined in the following way:

> If **P** is true and **Q** is true, then **P** ↔ **Q** is true.
> If **P** is true and **Q** is false, then **P** ↔ **Q** is false.
> If **P** is false and **Q** is true, then **P** ↔ **Q** is false.
> If **P** is false and **Q** is false, then **P** ↔ **Q** is true.

E X E R C I S E 5

Tell whether **P** ↔ **Q** is true (T) or false (F) in each of the following cases:

A. 1. Where **P** is true and **Q** is false.
  2. Where both **P** and **Q** are true.
  3. Where **P** is false and **Q** is true.
  4. Where **P** is true and **Q** is true.
  5. Where **P** is false and **Q** is false.

B. Is **P** ↔ ¬**P** always true, always false, or does the truth value of this equivalence depend on the truth value of **P**?

▶ 3.3 *Diagrams of Truth Value*

No matter how long and complicated a molecular sentence may be, we can work out its truth value if we know the truth values of its parts. We do this for each type of molecular sentence by using the rules of usage which were explained in the sections preceding this one. One way to analyze the truth value of a molecular sentence is to diagram it. Suppose we have the sentence: (**P** ∨ **Q**) & **R**. Suppose also that **P** is a true sentence, **Q** is a false sentence, and **R** is a true sentence. The diagram will then have the form,

$$(P \lor Q) \ \& \ R$$

$$\text{T} \quad \text{F} \quad\quad \text{T}$$

We begin with the atomic sentences, then the smallest molecular sentence, in this case **P** ∨ **Q**, and work out until the final loop connects the two members of the major connective. For an atomic sentence the 'T' or 'F' is put directly under the atomic letter. For a molecular sentence the 'T' or 'F' is put under the connective that dominates that sentence. The major connective in the example is 'and'. Since P is true, the sentence P ∨ Q is true according to our rule of usage for disjunctions. Therefore we write the letter 'T' in the loop which connects the two members of the disjunction. In order that the dominant connective, the conjunction, be a true sentence both its parts must be true. We see that this is the case because P ∨ Q is true and the other side of the conjunction R is also true. The big loop connecting the major connective has a 'T' for true. This means that the entire sentence is a true sentence.

Let us consider an example,

$$(P \ \& \ Q \rightarrow P) \ \& \ (R \lor S)$$

where P is true, Q is true, R is false, and S is false. (Before continuing, you should be cautioned that if an atomic sentence occurs more than once within the entire molecular sentence then it must be treated *the same way* each time it occurs. This means that if sentence P is true in one part of a molecular sentence, then it must be true each time it occurs in that sentence. If sentence Q is false, it must be false wherever it appears. In our example above, sentence P occurs twice. Whatever its truth value, it must have the same truth value in both occurrences.)

The diagram of our example will have the form

We begin with the smallest molecular sentences and work from the inside out. The first loops are for the conjunction and the disjunction; then we can add the loop for the conditional sentence, P & Q → P. We end with the loop connecting the two parts of the conjunction since the conjunction is the major connective. As it turned out, the left-hand side of the conjunction is true but the right-hand side is false. Therefore the conjunction itself, our complete molecular sentence, is a false sentence. Notice that the sentence P is true in both occurrences, as required. Our final example will be a still more complicated molecular sentence. It is

$$[(A \lor B) \And \neg A] \to \neg(C \to A).$$

In this sentence let **A** be true, **B** be false, and **C** be false. The diagram below shows that the entire sentence, which is a conditional, is a true sentence.

$$[(A \lor B) \And \neg A] \to \neg(C \to A).$$

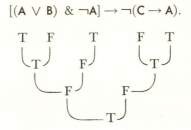

Notice the sentence that is a negation. Since **A** is a true sentence then ¬**A** is false. Also notice that the dominant connective is →, and therefore its loop is drawn last. In the case of the conditional the antecedent was false and the consequent was true. The rule of usage for conditionals tells us that in this case the conditional is a true sentence.

The exercise below asks you to diagram the truth of some molecular sentences. If you are not certain about the rules of usage for some of the connectives you can go back to the section on those connectives and review the rules.

EXERCISE 6

A.   Show the truth values of each of the following sentences by diagramming them, if **P** and **Q** are true sentences and **A** and **B** are false sentences.

1. P → (P → Q)
2. (A → P) → (P → A)
3. (P → A) → (A → P)
4. (P → A) → (¬P → ¬A)
5. ¬(P & Q) → (¬P ∨ ¬Q)
6. ¬(P & B) → (¬P & ¬B)
7. [(P & Q) → B] → [P → (Q → B)]
8. ¬[(P ∨ B) & (B ∨ A)]
9. (¬P ∨ B) ∨ (¬B & A)
10. (P → Q) ↔ (Q → P)

B.   Which of the following sentences are true, if we let

> **N**='New York is larger than Chicago'
> **W**='New York is north of Washington'
> **C**='Chicago is larger than New York'.
> (Thus **N** and **W** are true and **C** is false.)?

1. N ∨ C
2. N & C
3. ¬N & ¬C
4. N ↔ ¬W ∨ C
5. W ∨ ¬C → N
6. (W ∨ N) → (W → ¬C)
7. (W ↔ ¬N) ↔ (N ↔ C)
8. (W → N) → [(N → ¬C) → (¬C → W)]

C.   Let

$$P='2+4=6'$$
$$Q='2+8=10'$$
$$R='3\times4=12'$$
$$S='2\times0=2'.$$

You can tell the truth value of **P**, **Q**, **R**, and **S**. Find the truth values of the following sentences:

1. (P & Q) & (R & S) → P ∨ S
2. P & Q ↔ R & ¬S
3. (P → Q) → [(Q → R) → (R → S)]
4. (P ↔ Q) → (S ↔ R)
5. (P & Q) ∨ S → (P ↔ S)
6. (P → ¬Q) ↔ (P ∨ R) & S
7. (Q & R) & S → (P ↔ S)
8. (¬P → Q) → (S → R)
9. S → P & Q
10. P & Q → S

D.   Let '$x=0$' and '$x=y$' be true, and let '$y=z$' and '$y=w$' be false. Find the truth values of the following sentences.

1. If $x=0$ and $x=y$ then $y\neq z$.
2. If $x\neq0$ or $y=w$ then $y=z$.
3. If $x\neq y$ or $y\neq z$ then $y=w$.
4. If $x\neq0$ or $x\neq y$ then $y\neq z$.

5. If $x=0$ then $x \neq y$ or $y \neq w$.
6. If $x \neq 0$ then $y = z$.

▶ 3.4 *Invalid Conclusions*

So far all of the exercises in this book that ask you to derive conclusions from sets of premises have been those in which the suggested conclusion was actually valid; it did follow from the premises given. Logic would be a trivial subject indeed if we always knew in advance that the conclusion followed from the premises. Of course, this is not the case. We must be prepared to deal with situations in which we do not know whether the particular conclusion follows or does not follow from the given premises. We shall want to be able to prove that a conclusion does not follow logically and that a particular inference is *invalid* (not valid).

Suppose we are given a set of premises and asked to prove that a certain conclusion follows logically from these premises but we are unable to derive the desired conclusion. We cannot assume that the conclusion is therefore invalid or does not follow from the premises. There must be some method of proving beyond doubt that a conclusion does not follow from the given premises. An example of such a proof is given below.

The following argument is invalid:

> If you are his son then he is your parent.
> He is your parent.
> Therefore, you are his son.

To say that this argument is invalid is to say that the conclusion does not logically follow from the premises. We are referring to the *form* of the argument when we talk about valid or invalid conclusions. The argument is either valid or invalid according to its logical form. If we symbolize the argument, the form itself will be easier to see. Symbolized, the argument would appear as

$$P \rightarrow Q$$
$$\underline{\quad Q \quad}$$
$$P$$

If this were a valid form, it would *always* permit only true conclusions to be derived from true premises. Therefore, if there is any case in which this form permits a false conclusion to be derived from premises that are true, then it cannot be valid. To prove an argument is invalid we find an *interpretation* of the argument where the premises are true sentences and the conclusion is a false sentence. We can interpret the argument by replacing its different atomic sentences by any sentences we choose. The form must remain the same.*

To prove that the argument above is invalid, we could interpret it in the following way:

Let

P = 'You are a citizen of Maine'.

Q = 'You are a citizen of the United States'.

The interpretation will read

If you are a citizen of Maine, then you are a citizen of the United States.
You are a citizen of the United States.
Therefore, you are a citizen of Maine.

There are certainly many cases in which these premises are true sentences but the conclusion is false. For any citizen of the United States the premises are true but for most citizens of the United States the conclusion is false. The form of the original argument permitted us to derive a false conclusion from true premises. Therefore, we have shown that the argument is not valid.

This argument we have been considering is an example of a common fallacy which is known as the Fallacy of Affirming the Consequent. What was important in this interpretation was not the content of the sentences 'you are a citizen of Maine' and 'you are a citizen of the United States', but their possible truth values.

To show that an inference is invalid we may therefore give an interpretation in terms of truth values and not consider any particular sentences. There are two main steps in showing a conclusion is invalid or that an argument is fallacious.

(1) Symbolize the premises and conclusion.
(2) Find an assignment of truth values for the atomic sentences such that *all* the premises are true and the conclusion is false.

---

* Each atomic sentence must have the same interpretation every time it occurs.

The analysis of the Fallacy of Affirming the Consequent just discussed is

P → Q		T	F
Q		Q	P
P			

Premises P → Q	Q	conclusion P
F T	T	F

⌐T⌐   invalid

Notice that four things are shown: (1) the symbolized premises and conclusions, (2) the assignment of truth values for the atomic sentences, (3) a truth diagram of each premise, and (4) a truth diagram of the conclusion. In order to show the conclusion is invalid the truth diagrams must have the value T for every premise and F for the conclusion.

This method of assigning T or F to the atomic sentences to show that a conclusion is invalid we call the *method of truth assignment*.

Another common fallacy is the Fallacy of Denying the Antecedent. We can use the method of truth assignment to show the invalidity of this form of inference. An example of the fallacy is the following:

> If this is Saturday, then tomorrow is Sunday.
> This is not Saturday.
> Therefore, tomorrow is not Sunday.

The conclusion is, *in fact*, a true one when the second premise is true. Yet the inference itself is not valid. Remember that a valid inference is such that the form permits *only* true conclusions to be derived from true premises. The fact that the conclusion is true in the example above does not prove the validity of the inference. The form of the argument is

$$P \rightarrow Q$$
$$\neg P$$
$$\overline{\neg Q}$$

It is possible to find truth assignments such that the premises are true but the conclusion false. For example,

P → Q		T	F
¬P		Q	P
¬Q			

Premises P → Q   ¬P   conclusion ¬Q

    F   T   F   T

      T   T   F

                                          invalid

The truth diagrams show that assigning F to P and T to Q makes both premises true but the conclusion false. Can you think of an example in English sentences of this form (Fallacy of Denying the Antecedent) in which premises are true and conclusion is false?

We consider a more complicated example. We want to show that the conclusion given below the line does not logically follow from the two premises.

(P & Q) → (P → R) ∨ S   T | F
P & ¬R

                                  P | R
¬P ∨ ¬Q   Q

                                         S

Premises P & Q → (P → R) ∨ S   P & ¬R

    T   T   T   F   T   T   F

      T   F   T

                         T   T

          T

conclusion ¬P ∨ ¬Q

      T   T

    F   F

        F   invalid

EXERCISE 7

A. The following arguments are invalid. For each argument, give a truth assignment that will prove its invalidity.

1. If Mary leaves early, then she walks home with Sally.
Either she walks home with Sally or she meets Karen.
Mary leaves early.
Therefore, she does not meet Karen.

2. Either the water is cold or the day is not hot.
   The day is hot. '
   If the pool has just been filled, then the water is cold.
   Therefore, the pool has just been filled.
3. George is elected if and only if the voting is heavy.
   The voting is heavy.
   Either George is elected or Sam will not be appointed.
   Therefore, Sam will be appointed.
4. If Sue is chosen winner, then Jan is out of the contest.
   If Sue is chosen winner, then Sandra is out of the contest too.
   Jan is out of the contest and Sandra is out of the contest too.
   Therefore, Sue is chosen winner.
5. Either the animal is not a bird or it has wings.
   If the animal is a bird, then it lays eggs.
   The animal does not have wings.
   Therefore, it does not lay eggs.
6. Either subtraction is not always possible in the number system or the system includes more than the natural numbers.
   If subtraction is always possible in the number system, then the system includes the negative integers.
   The system does not include more than the natural numbers.
   Therefore, the system does not include the negative integers.

B. If any of the symbolized arguments below are valid, show a derivation of the conclusion by means of a complete formal proof. If they are invalid, prove it by a truth assignment.

1. Prove: $\neg$S
   (1) T & S $\leftrightarrow$ R
   (2) $\neg$R
   (3) T

2. Prove: S
   (1) Q $\rightarrow$ R
   (2) P $\rightarrow$ Q
   (3) P $\vee$ T
   (4) T $\rightarrow$ S
   (5) $\neg$R

3. Prove: $\neg$Q
   (1) T $\rightarrow$ Q
   (2) $\neg$T $\vee$ R
   (3) $\neg$R

4. Prove: S
   (1) R $\vee$ S
   (2) $\neg$P
   (3) Q $\vee$ $\neg$R
   (4) P $\leftrightarrow$ Q

5. Prove: T
    (1) ¬(P ∨ Q)
    (2) P ∨ R
    (3) T → R

6. Prove: ¬R
    (1) P → T
    (2) Q → S
    (3) S ∨ R
    (4) P ∨ ¬Q

7. Prove: ¬S
    (1) ¬(P & R)
    (2) Q → R
    (3) Q ∨ ¬S

8. Prove: ¬P
    (1) Q → R
    (2) ¬R → S
    (3) ¬T ∨ ¬P
    (4) (Q → S) → T

9. Prove: R ∨ ¬Q
    (1) S & ¬T
    (2) T → P
    (3) S → R
    (4) ¬P → ¬Q

10. Prove: ¬T
    (1) ¬P
    (2) ¬Q ∨ ¬R
    (3) Q ↔ P
    (4) T → R

C. Show by a formal derivation or a truth assignment whether each of the following arguments is valid or invalid:

1. Either John and Jim are the same age, or John is older than Jim.
   If John and Jim are the same age, then Pam and John are not the same age.
   If John is older than Jim, then John is older than Mary.
   Therefore, either Pam and John are not the same age or John is older than Mary.

2. If this is December, then last month was November.
   If last month was November, then six months ago it was June.
   If six months ago it was June, then eleven months ago it was January.
   If next month will be January, then this is December.
   Last month was November.
   Therefore, this is December.

3. If the contract is valid, then Jones will not lose the lawsuit.
   If Jones loses the lawsuit, then he must pay damages.
   If he must pay damages, then Smith will not get his money.
   Therefore, either Smith will not get his money or the contract is not valid.

4. If Mary is correct, then Jack is wrong.
   If Jack is wrong, then Terry is mistaken too.
   If Terry is mistaken too, then the show is not tonight.
   Either the show is tonight or Bob will not see it.
   Mary is correct.
   Therefore, Bob will not see the show.

5. If Brown performed the contract, then the goods were delivered on time.
   Brown either performed the contract or his records are in error.
   If his records are in error, then he did not order delivery on the seventh.
   Therefore, the goods were not delivered on time.

6. $x^2 = 9 \quad \rightarrow \quad x = 3 \quad \vee \quad x = -3$
   $x = 3 \quad \vee \quad x = -3 \quad \rightarrow \quad xy < 20$
   $xy \not< 20$
   Therefore: $x^2 = 9 \quad \vee \quad xy < 20$

7. $x^2 = 9 \quad \rightarrow \quad x = 3 \quad \vee \quad x = -3$
   $x = 3 \quad \vee \quad x = -3 \quad \rightarrow \quad xy < 20$
   $xy \not< 20$
   Therefore: $x^2 \neq 9$

8. $x \neq 0$
   $x = 0 \quad \vee \quad \neg(x < 1 \quad \vee \quad y \not> x)$
   $y > x \quad \rightarrow \quad y > 1 \quad \& \quad x + y > 2$
   Therefore: $y > 1 \quad \rightarrow \quad x < 1$

9. $x \neq 0$
   $x = 0 \quad \vee \quad \neg(x < 1 \quad \vee \quad y \not> x)$
   $y > x \quad \rightarrow \quad y > 1 \quad \& \quad x + y > 2$
   Therefore: $x + y > 2 \quad \& \quad y > 1$

10. $x^2 - 3x + 2 = 0 \quad \rightarrow \quad x = 1 \quad \vee \quad x = 2$
    $x = 1 \quad \vee \quad x = 2 \quad \rightarrow \quad 3x > x^2$
    $3x \not> x^2$
    Therefore: $3x > x^2 \quad \vee \quad x = 1$

D. In the following derivations there are several errors. Find the errors and correct them so that the formal proofs will be perfectly correct.

1. (1) P   P   2. (1) P & Q   P
   (2) ¬T ∨ ¬Q   P   (2) P → ¬R   P
   (3) ¬Q → ¬P   P   (3) Q → ¬S   P
   (4) ¬¬P   DN 1   (4) P   S 1
   (5) ¬Q   TT 3, 4   (5) ¬R   TT 2, 4
   (6) ¬T   TT 2, 6   (6) Q   A 1
                                      (7) ¬S   PP 2, 6
                                      (8) ¬R & S   A 5, 7
                                      (9) ¬(R ∨ S)   DS 8

        3. (1) R → Q   P
           (2) P → Q   P
           (3) P ∨ R   P
           (4) T & S   P
           (5) Q ∨ Q   DS 1, 2, 3
           (6) Q   DN 5
           (7) S   A 4
           (8) Q & S   A 5, 6
           (9) (Q & S) ∨ U   DS 8

▶ 3.5 *Conditional Proof*

At this point in our study of logic we are able to do some complicated proofs. You will also find that there are some very simple derivations which are not yet possible with just the list of rules introduced. An example of a fairly obvious conclusion that we cannot yet derive is the following:

> If Bill wins then Joe is second.
> If Karl is second then Joe is not second.
> Therefore, if Karl is second then Bill does not win.

Let us symbolize this argument to decide whether or not we are able to prove that it is valid.

Let

$$P = \text{'Bill wins'}$$
$$Q = \text{'Joe is second'}$$
$$R = \text{'Karl is second'}.$$

The symbolization of the entire argument is

$$P \rightarrow Q$$
$$R \rightarrow \neg Q$$
$$\overline{R \rightarrow \neg P}$$

The rules that we know are not sufficient to derive the conclusion in this argument. Yet it would be impossible to find a truth assignment making the premises true and the conclusion false. To make the conclusion false both **R** and **P** must be true. But then one or the other of the premises is false no matter which truth value assignment is given **Q**. To derive the conclusion, which is valid, we will need a rule which has not yet been introduced.

The *Rule of Premises*, rule P, permits us to introduce a new premise wherever we wish in a proof. This may be any sentence we choose. This may seem absurd at first because it would appear that if we are allowed to introduce any premise at any time then eventually we could prove anything at all. The point is, of course, that any logical argument rests upon *all* the premises it uses. If we introduce a new premise then whatever conclusion we derive from the full set of premises will rest upon that full set and not just upon the original set of premises. In other words, any logically correct argument is no better or worse than the premises on which it rests. We are not able to use rule P to derive just any conclusion from a *given* set of premises because at the moment we introduce the new premise, any sentence then derived rests upon *all* the premises we used, including the new premise.

In order to show the complete set of premises on which a conclusion is based, we shall use the following method. Whenever a new premise is introduced in a derivation we will immediately move the entire proof a few spaces to the right. This will indicate that whatever sentence is derived in this right-hand portion of the formal proof depends not only upon the original set of given premises but also upon the additional premise introduced.

Let us use the example given above to show a case of the new use of the rule P. In line (3) a new premise R is introduced. From this line onward note that the proof is moved several spaces to the right.

(1)	$P \rightarrow Q$		P
(2)	$R \rightarrow \neg Q$		P
(3)		R	P
(4)		$\neg Q$	PP 2, 3
(5)		$\neg P$	TT 1, 4

In the foregoing derivation notice the capital letter 'P' after the sentence **R** in the third line. This shows that adding sentence **R** is justified by the rules of premises, rule P. And moving it several spaces to the right shows that **R** is not one of the original premises. Using **R**, we were able to derive the sentence ¬**P**. The sentence ¬**P**, however, does not rest upon our original set of premises alone but upon the new set of premises formed by adding sentence **R**. It is essential to realize that ¬**P** is not derived from the original argument itself and does not follow logically from sentences **P** → **Q** and **R** → ¬**Q**.

We might have summarized the whole idea in this derivation if we had said that from our original premises, *if* we also have **R** then we can get ¬**P**. This comes very close to the idea of the new rule we will introduce, a rule which will permit us to complete the derivation of the conclusion in the original argument. If you look at that argument you will see that we are trying to prove the conditional sentence **R** → ¬**P**.

This new rule, the *Rule of Conditional Proof* (CP) is summarized:

If we are able to derive a sentence **S** *from a sentence* **R** *and a set of premises, then we may derive* **R** → **S** *from the set of premises alone.*

In the derivation above we were able to derive the sentence ¬**P** from our added premises **R** plus the original set of premises. Therefore, we may derive the conditional sentence **R** → ¬**P** from the set of premises alone. Since the conditional sentence **R** → ¬**P** is derived from the original set of given premises alone we move this line of proof to the left in order to put it back in line with the original premises.

The complete derivation of the conclusion from the premises in our example will now appear as

(1) P → Q		P
(2) R → ¬Q		P
(3)	R	P
(4)	¬Q	PP 2, 3
(5)	¬P	TT 1, 4
(6) R → ¬P		CP 3, 5

Notice in the complete derivation above that we derive line (6) by means of conditional proof using lines (3) and (5). Line (3) is the line in which we introduced the antecedent of the conditional and line (5) is the line in which we were able to derive the consequent of the conditional sentence. Notice also that line (6) is moved back in line with the original premises. This is because line (6) rests upon the original set of premises alone.

The intuitive idea of a conditional proof is really quite simple. The desired conclusion is a conditional sentence, with the 'if . . . then . . .' connective. In the example above, the conclusion we wish to reach said 'if R then ¬P'. In asking whether the conclusion followed from the given premises, we are asking, in effect, whether with the premises given, *if* we have R then can we get ¬P. In line (3) we are saying, "Let us add R and see". Line (5) shows us that if we did have R and our set of original premises we would then have ¬P. Therefore, we know that from our original set of premises we may derive 'if R then ¬P'.

A good strategy to follow is this: If the desired conclusion of an argument is a conditional sentence, add the antecedent as a new premise, move the proof several spaces to the right and try to derive the consequent from the original set of premises plus the added premise. If you can derive the consequent by adding the antecedent as a premise then by rule CP you can prove that the conditional sentence follows from the original set of premises so the proof is moved back to the left. It is not always necessary to use a conditional proof to derive a conditional sentence as a conclusion. But if you cannot see another possible derivation it is well to introduce the antecedent as a new premise and try for a conditional proof.

Another example of the use of conditional proof in a derivation is given below. The conclusion desired is the sentence D → C.

(1)	A → (B → C)		P
(2)	¬D ∨ A		P
(3)	B		P
(4)		D	P
(5)		A	TP 2, 4
(6)		B → C	PP 1, 5
(7)		C	PP 3, 6
(8)	D → C		CP 4, 7

The portion of the proof that is moved several spaces to the right (or indented) is called a *subordinate* proof. It gives an answer to the question, "What could be proved if we also had the premise D in addition to what we have so far?" In the subordinate proof we in effect say, "Let us see!" And we find that if we had D then we could get C. So in line (8) we say that with the original premises, if D then C.

Now rule P says that a premise may be introduced at *any* point in a derivation. And our new requirement is that *whenever* this is done for a new premise in addition to the given premises then the proof should be moved to the right. Suppose we are already indented in a subordinate proof. Study the following example.

$$\text{Prove}: P \rightarrow (\neg Q \rightarrow R)$$

(1)	S & (¬P ∨ M)		P
(2)	M → Q ∨ R		P
(3)	¬P ∨ M		S 1
(4)	P		P
(5)	M		TP 3, 4
(6)		¬Q	P
(7)		Q ∨ R	PP 2, 5
(8)		R	TP 6, 7
(9)	¬Q → R		CP 7, 8
(10)	P → (¬Q → R)		CP 4, 9

The conclusion is to be a conditional so a conditional proof is tried.

We introduce the antecedent, P, of the desired conditional, and try to derive the consequent, ¬Q → R. In line (6), we try another conditional proof since the ¬Q → R we are trying to derive is itself a conditional. So we add its antecedent ¬Q and indent a second time. This gives a proof subordinate to the first subordinate proof. After deriving R, the rule for conditional proof is used. This allows a return to the proof to which we are subordinate and gives the consequent we were trying to derive in the first subordinate proof. Applying CP again brings us back to the main proof. One application of CP ends only one indentation, that is, only one subordination. One other thing needs emphasis: at any stage any line may be used which appears earlier in the *same* proof or earlier in any proof to which we are subordinate. Thus in line (7) we may use line (2) and line (5). But after line (9) we could not use lines (6) to (8) and after line (10) we could not use lines (4) to (9).

EXERCISE 8

A. Use a conditional proof to derive the conclusion called for in each of the following symbolized arguments. Show the complete formal proof.

1. Prove: $\neg P \rightarrow Q$
   (1) $P \vee Q$

2. Prove: $R \rightarrow \neg Q$
   (1) $\neg R \vee \neg S$
   (2) $Q \rightarrow S$

3. Prove: $C \rightarrow \neg D$
   (1) $B \rightarrow \neg C$
   (2) $\neg(D \;\&\; \neg B)$

4. Prove: $\neg Q \rightarrow T$
   (1) $S \rightarrow R$
   (2) $S \vee P$
   (3) $P \rightarrow Q$
   (4) $R \rightarrow T$

5. Prove: $P \rightarrow P \;\&\; Q$
   (1) $R \rightarrow T$
   (2) $T \rightarrow \neg S$
   (3) $(R \rightarrow \neg S) \rightarrow Q$

6. Prove: $S \rightarrow Q$
   (1) $R \rightarrow Q$
   (2) $T \rightarrow R$
   (3) $S \rightarrow T$

7. Prove: $\neg(R \;\&\; S) \rightarrow T$
   (1) $\neg P$
   (2) $\neg R \rightarrow T$
   (3) $\neg S \rightarrow P$

8. Prove: $T \vee \neg S \rightarrow R$
   (1) $\neg R \rightarrow Q$
   (2) $T \rightarrow \neg Q$
   (3) $\neg S \rightarrow \neg Q$

9. Prove: $T \rightarrow \neg(P \vee Q)$
   (1) $\neg S \vee \neg P$
   (2) $Q \rightarrow \neg R$
   (3) $T \rightarrow S \;\&\; R$

10. Prove: $\neg Q \rightarrow T \;\&\; S$
    (1) $R \rightarrow S$
    (2) $S \rightarrow Q$
    (3) $R \vee (S \;\&\; T)$

11. Prove: $(P \;\&\; Q) \rightarrow (S \;\&\; T)$
    (1) $R \vee S$
    (2) $\neg T \rightarrow \neg P$
    (3) $R \rightarrow \neg Q$

12. Prove: $S \rightarrow P \vee Q$
    (1) $S \rightarrow T$
    (2) $R \rightarrow P$
    (3) $T \rightarrow R$

13. Prove: $\neg(P \vee R) \rightarrow T$
    (1) $Q \rightarrow P$
    (2) $T \vee S$
    (3) $Q \vee \neg S$

14. Prove: $E \rightarrow K$
    (1) $E \vee F \rightarrow G$
    (2) $J \rightarrow \neg G \;\&\; \neg H$
    (3) $J \vee K$

15. Prove: $Q \leftrightarrow \neg P$
    (1) $\neg(\neg P \;\&\; \neg Q)$
    (2) $S \rightarrow \neg Q$
    (3) $\neg P \vee S$

16. Prove: $P \rightarrow (Q \rightarrow R)$
    (1) $P \;\&\; Q \rightarrow R$

17. Prove: $x=0 \;\; \vee \;\; x=1 \;\; \rightarrow \;\; x^3 - 3x^2 + 2x = 0$
    (1) $x=0 \;\; \rightarrow \;\; x^2 - x = 0$
    (2) $x=1 \;\; \rightarrow \;\; x^2 - x = 0$
    (3) $x=2 \;\; \vee \;\; x^2 - x = 0 \;\; \rightarrow \;\; x^3 - 3x^2 + 2x = 0$

18. Prove: $y=2 \lor y=4 \to y<4 \lor y>3$
   (1) $(y=4 \leftrightarrow x>y) \ \& \ x>z$
   (2) $x>y \lor z>y \to y<4 \ \& \ y\neq3$
   (3) $y=2 \to z>y$

19. Prove: $y=2 \to x=y$
   (1) $x\neq y \to x>y \lor y>x$
   (2) $y\neq2 \lor x=2$
   (3) $x>y \lor y>x \to x\neq2$

20. Prove: $x=1 \to x\neq2 \ \& \ y\neq1$
   (1) $x=1 \to xy=2$
   (2) $x+y\neq3 \to x\neq1$
   (3) $y=1 \lor x=2 \to \neg(x+y=3 \ \& \ xy=2)$

B. Show a complete derivation for each of the following arguments to prove that it is valid.

1. Either the witness is not telling the truth or Brown was home by eleven.
   If Brown was home by eleven then he saw his uncle.
   If he saw his uncle then he knows who was there earlier.
   Therefore, if the witness is telling the truth then Brown knows who was there earlier.

2. Either logic is difficult or not many students like it.
   If mathematics is easy, then logic is not difficult.
   Therefore, if many students like logic, mathematics is not easy.

3. If the Pirates are third, then if the Dodgers are second the Braves will be fifth.
   Either the Giants will not be first or the Pirates will be third.
   In fact, the Dodgers will be second.
   Therefore, if the Giants are first, then the Braves will be fifth.

4. If Joe wins, then either Ron or Steve will be second.
   If Ron is second then Joe will not win.
   If Mike is second then Steve will not be second.
   Therefore, if Joe wins, then Mike will not be second.

5. If Betty is her sister then Charles is her brother.
   If Charles is her brother then she lives on West Fourth Street.
   Therefore, if Betty is her sister then she lives on West Fourth Street.

C. If the following arguments are valid, show a formal proof. If they are invalid, write 'invalid' beside them and prove invalidity by a truth assignment.

    1. If Ken is not first then Bill is first.
       But Bill is not first.
       Either Ken is first or Paul is third.
       If Sam is second then Paul is not third.
       Therefore, Sam is not second.

    2. If the contract is legal and if Smith entered into the contract then Jones will win the lawsuit.
       Either Jones will not win the lawsuit or Smith will be liable.
       Smith will not be liable.
       Therefore, either the contract is not legal or Smith did not enter into the contract.

    3. If we wait for Sue then we will be late.
       Either we will not be late or we will reach school after 8:30.
       If we reach school after 8:30 then we check in at the office.
       Therefore, if we wait for Sue then we either check in at the office or have a written excuse.

    4. Marty was appointed chairman.
       If Ted was elected then Marty was not appointed chairman.
       If James was elected then the selection was made today.
       Therefore, if either Ted was elected or James was elected then the selection was made today.

    5. Either Mary is with Sally or Jane is with Sally.
       If today is Monday then Mary is with Sally.
       Today is Monday.
       Therefore, Jane is not with Sally.

D. Find the errors in the following derivations and correct them.

    1. (1) $\neg T \lor \neg R$ P
       (2) $S \rightarrow T \,\&\, R$ P
       (3) $Q \rightarrow S$ P
       (4) $Q \lor P$ P
       (5) $\neg(T \,\&\, R)$ DL 1
       (6) $\neg S$ TP 2, 3
       (7) $\neg\neg Q$ TT 3, 6
       (8) P TP 3, 7

2. (1) ¬S → ¬T   P
   (2) T   P
   (3) S → R & Q   P
   (4) Q & R → P   P
   (5) ¬S   TT 1, 2
   (6) S   DN 5
   (7) R & Q   TT 3, 6
   (8) Q & R   A 7
   (9) P   PP 5, 8

3. (1) T → ¬R   P
   (2) S → Q   P
   (3) ¬Q ∨ R   P
   (4) T   P
   (5)   ¬R   PP 2, 4
   (6)   Q   TP 3, 5
   (7)   ¬S   PP 2, 6
   (8) ¬S   CP 4, 7

E.   If the following arguments are valid, show a formal proof. If they are invalid, write 'invalid' beside them and prove invalidity by a truth assignment.

1. If $x=0$ then $x+y=y$
   If $y=z$ then $x+y\neq y$
   Therefore, if $x=0$ then $y\neq z$.
2. If $x=0$, then $y\not<z$
   Either $x\neq 0$ or $x<y$.
   $y<z$
   If $z<w$ then $x\not<y$
   Therefore, $z\not<w$.
3. If $x<y$ then $y=z$
   If $x>y$ then $y>z$
   Therefore, if $x<y$ or $x>y$ then $y=z$ or $y>z$.
4. If $x=y$ then $y=z$
   If $y=z$ then $x=0$
   If $x\neq 0$ then $w\neq 0$.
   Therefore, if $w=0$ then $x\neq y$.
5. Either $x>y$ or $y>x$.
   If $y>x$ then $x\neq 0$
   If $x\neq 0$ then $y\neq w$
   Therefore, if $y=w$ then $x>y$.

6. If $y=0$ then $x>z$.
   If $x>z$ then $z=w$.
   Either $y=0$, or $x>z$ and $x=0$
   Therefore, if $z\neq w$, then $x=0$ and $x>z$.

▶ 3.6 *Consistency*

Look at the following three sentences:

1. Richard Nixon won the Presidential election of 1960.
2. My little brother can lift a two-ton object.
3. Tom is taller than Bill and Tom is not taller than Bill.

You will probably agree that each of the sentences is false. We know that they are false for different reasons, however.

The first sentence, we know to be false in *fact*. But it could, under different circumstances, have been true. We would also say that the second sentence is false, because no boy is in fact strong enough to lift that weight under ordinary conditions. Still we could suggest circumstances where the sentence might be true; for example, in a space ship where everything is weightless.

The third sentence, however, *could not possibly be true*, not ever or anywhere. It is a *logically* impossible sentence. It is not even necessary that we know the meaning of 'Tom is taller than Bill' to know that it cannot be true. We know this by its logical *form*.

A sentence of the logical form

$$(R) \ \& \ \neg(R)$$

is called a *contradiction*. We say that two sentences are contradictory if one is the negation of the other. A contradiction, then, is the conjunction of a sentence and its negation. It is always false. The sentence (3) above, is of the form

$$R \ \& \ \neg R$$

and therefore it is a contradiction.

With two truth diagrams we can easily show that a contradiction is logically false.

$$R \ \& \ \neg R$$

T   T

F

F

$$R \ \& \ \neg R$$

F   F

T

F

They show that **R** & **¬R** is false no matter what the truth value of the atomic sentence **R**. There is no possibility of its being true. This is the reason it is called *logically* false.

There are many forms a sentence may take which make it logically false. If the sentence is not of the form of a contradiction (**R** & **¬R**) it can be recognized by using the rules of derivation to derive a contradiction. The rules never allow us to derive a false conclusion from a *true* premise, so if the conclusion is logically false then the premise must be logically false also. This is illustrated by

Example a.

$$(1) \quad \neg(\neg S \ \lor \ S) \qquad P$$

This sentence is not in the form of a contradiction but we can derive

$$\neg\neg S \ \& \ \neg S \qquad DL$$

and

$$\neg S \ \& \ \neg\neg S \qquad CL.$$

Line (3) is in the form **R** & **¬R** and this proves that the premise cannot be true.

Another example in which the premise is logically false is

Example b.

$$(1) \quad (S \to R) \ \& \ \neg(\neg S \ \lor \ R) \qquad P$$
$$(2) \quad S \to R \qquad S \ 1$$
$$(3) \quad \neg(\neg S \ \lor \ R) \qquad S \ 1$$
$$(4) \quad S \ \& \ \neg R \qquad DL \ 3$$
$$(5) \quad S \qquad S \ 4$$
$$(6) \quad \neg R \qquad S \ 4$$
$$(7) \quad R \qquad PP \ 2, 5$$
$$(8) \quad R \ \& \ \neg R \qquad A \ 7, 6$$

The last two lines of this proof could just as well be

(7) ¬S	TT 2, 6
(8) S & ¬S	A 5, 7.

It does not matter what particular contradiction is derived.

EXERCISE 9

Symbolize each of the following sentences. Tell whether or not it is logically false. If it is, derive a contradiction to prove it.

1. This is Thursday or this is not Thursday.
2. If Jane is tall then her brother is not short, but Jane is tall and her brother is short.*
3. It is not the case that either Jim won his match or Jim did not win his match.
4. Both A is not equal to B and C is equal to B, and if C is equal to B then A is equal to B.
5. $\neg(\neg(x<2 \ \lor \ x=2) \ \lor \ \neg(x \not< 2 \ \& \ x \neq 2))$

We know that a contradiction, **P** & ¬**P**, is the conjunction of a sentence and its negation. If one of the sentences in the conjunction is a true sentence then the other must be false; they logically cannot both be true. Their conjunction, then, must be a false sentence. Any two or more sentences that logically cannot all be true together are said to be *inconsistent*. We say that they form an inconsistent set of sentences and together they imply a contradiction.

In some cases, we shall be interested not in deriving a particular conclusion but in deciding whether a set of premises is consistent or inconsistent. To prove premises inconsistent we derive a contradiction. Using the rules and methods of derivation we already know, if we are able to derive a contradiction from the premises then this is proof that the set of premises is inconsistent. We have proved that they cannot all be true together.

We go about the proof in the same way we go about deriving a conclusion. In this case, however, we are not trying to derive some

---

* In this sentence as in most sentences 'but' is logically equivalent to 'and'.

particular conclusion but just any contradiction. It does not matter what contradiction is derived. It can be any sentence which has the form, P & ¬P.

Example c.

(1) If Dick wins the race then Jack places second.
(2) Dick wins the race.
(3) Jack does not place second.

You can see that these three sentences cannot all three be true at once. Yet any two of them could be true. They may be translated:

(1) D → J
(2) D
(3) ¬J

Either of the following proofs shows that they are inconsistent.

(1) D → J	P	(1) D → J	P
(2) D	P	(2) D	P
(3) ¬J	P	(3) ¬J	P
(4) J	PP 1, 2	(4) ¬D	TT 1, 3
(5) J & ¬J	A 3, 4	(5) D & ¬D	A 2, 4

Example d.

(1) If the area is near the equator then the sun is almost always directly overhead.
If the sun is almost always directly overhead then the area has a hot tropical climate.
If the area is at a very high altitude then it does not have a hot tropical climate.
This area is near the equator and it is at a very high altitude.

The following derivation shows that it is possible to derive a contradiction (line 10) by using the rules of logical inference. Here is a proof that the set of premises in the example is inconsistent.

(1) E → S	P
(2) S → H	P
(3) A → ¬H	P
(4) E & A	P
(5) A	S 4
(6) ¬H	PP 3, 5
(7) E	S 4
(8) E → H	HS 1, 2
(9) H	PP 7, 8
(10) H & ¬H	A 6, 9

Another example of a derivation showing that the symbolized premises are inconsistent is

Example e.

(1) ¬(¬Q ∨ P)	P
(2) P ∨ ¬R	P
(3) Q → R	P
(4) ¬¬Q & ¬P	DL 1
(5) ¬¬Q	S 4
(6) Q	DN 5
(7) R	PP 3, 6
(8) ¬P	S 4
(9) ¬R	TP 2, 8
(10) R & ¬R	A 7, 9

EXERCISE 10

A. Prove that the following sets of premises are *inconsistent* by deriving a contradiction from each.

1. (1) ¬Q → R
   (2) ¬R ∨ S
   (3) ¬(P ∨ Q)
   (4) ¬P → ¬S

2. (1) T → P
   (2) T & R
   (3) Q → ¬R
   (4) P ∨ S → Q

3. (1) R → R & Q
   (2) ¬S ∨ R
   (3) ¬T ∨ ¬Q
   (4) S & T

4. (1) T ∨ ¬R
   (2) ¬(R → S)
   (3) T → S

5. (1) $Q \rightarrow P$
   (2) $\neg(P \lor R)$
   (3) $Q \lor R$

6. (1) $x=1 \rightarrow y<x$
   (2) $y<x \rightarrow y=0$
   (3) $\neg(y=0 \lor x\neq1)$

7. (1) $x=y \rightarrow x<4$
   (2) $x\not<4 \lor x<z$
   (3) $\neg(x<z \lor x\neq y)$

8. (1) $2\times5=5+5 \leftrightarrow 2\times6=6+6$
   (2) $3\times4=10 \leftrightarrow 4\times3=10$
   (3) $3\times4=10 \lor 2\times6=6+6$
   (4) $2\times5\neq5+5 \;\&\; 4\times3\neq10$

9. (1) $x<y \rightarrow x\neq y$
   (2) $y>z \rightarrow z\not<y$
   (3) $x=y \;\&\; y>z$
   (4) $x<y \lor z<y$

10. (1) $x=0 \leftrightarrow x+y=y$
    (2) $x>1 \;\&\; x=0$
    (3) $x+y=y \rightarrow x\not>1$

Suppose you do not know whether a given set of premises is consistent or inconsistent. You attempt to derive a contradiction from them and cannot. Your failure to prove inconsistency by deriving a contradiction is not itself a proof that the premises are consistent. You need a definite method or technique for proving the set of premises consistent. The method of truth assignment is the appropriate technique.

You will remember that when we say that premises are inconsistent we are saying that they cannot be true together. Therefore, if we are able to find at least one truth assignment in which all the premises are true together then we know that they are not inconsistent. A truth assignment for which all the premises are true gives a proof that those premises are consistent. Consider the premises:

> If Janet is a junior then Larry is a senior.
> If Larry is a senior then Martha is a junior.
> Martha is not a junior.

To prove them consistent, start by symbolizing the sentences to make their form clear. Then search for a truth assignment that will make all the sentences true.

	T	F
$J \rightarrow L$		J
$L \rightarrow M$		L
$\neg M$		M

Premises: J → L   L → M   ¬M

F F	F F	F
⌊ T ⌋	⌊ T ⌋	T ⌋

These diagrams demonstrate that it is possible for all the premises to be true under the one truth assignment for the atomic letters. This proves that the premises are consistent. The strategy we used in this example is the following. We note first that to make the premise ¬M true, we must assign F for false to M. Then to make the premise L → M true we must assign F to L because the consequent M has been assigned F for false. Having assigned F to L, to make the premise J → L true we must assign F to J, which completes our truth assignments for the atomic sentences. We then fill out the truth diagrams to demonstrate that this truth assignment does indeed make the premises all true together.

EXERCISE 11

A. Prove that the following sets of premises are *consistent* by showing interpretations in which all premises are true.

1. (1) Q & ¬S
   (2) ¬(P ∨ S)
   (3) Q → T

2. (1) P → Q
   (2) Q → R
   (3) ¬R ∨ S

3. (1) T → R
   (2) ¬R → S
   (3) S ∨ T

4. (1) ¬P ∨ ¬R
   (2) ¬P → S
   (3) ¬S

5. (1) R → Q
   (2) P → Q
   (3) Q → ¬T

6. (1) $3 \times 5 = 12$ → $6 + 8 = 11$
   (2) $6 + 8 = 11$ → $13 - 9 = 7$
   (3) $3 \times 5 \neq 12$ & $13 - 9 = 7$

B. For each of the following sets of premises, tell whether they are consistent or inconsistent. Prove your answer.

1. If Mary is oldest then Jerry is younger than Sue.
   Mary is oldest and Elizabeth is not older than Sue.
   It is not the case that either Sue is oldest or that Elizabeth is older than Sue.

2. A is the winner and C is not third.
   If A is the winner than B is fourth.
   If C is not third then B is not fourth.

3. John is in the library and it is not the case that either Tom is in history class or that Dick is in history class.
   If Chuck is in chemistry lab then Dick is in history class.
   If Ken is in geometry class then Tom is in history class.
   If John is in the library then either Chuck is in chemistry lab or Ken is in geometry class.

4. Jones is on the Council and either Smith will be elected or Black will be elected for next term.
   If Jones is on the Council then Black will not be elected for next term.
   If Smith will be elected then Jones will not serve out his complete term.

5. If Bill is six feet tall then it is not the case that both Steve is taller than Bill and Mark is taller than Bill.
   Bill is six feet tall and Mark is not as tall as Ned.
   If Mark is not taller than Bill then Mark is as tall as Ned.
   Furthermore, Steve is taller than Bill.

6. (1) A → C
   (2) ¬(B & C)
   (3) A ∨ (B & C)
   (4) B

7. (1) (B → C) → A
   (2) B → D
   (3) D → C
   (4) ¬A ∨ ¬D

8. (1) ¬(C ∨ D)
   (2) B ↔ C
   (3) C ↔ D
   (4) ¬B

9. (1) D ↔ ¬C
   (2) B → A
   (3) ¬(A ∨ ¬C)
   (4) D ∨ B

10. (1) B → A
    (2) B ∨ D
    (3) ¬(A & C)
    (4) ¬A ↔ ¬C

11. (1) $x=y \rightarrow x \neq y$
    (2) $x<y \lor x=y$
    (3) $x \nless y \rightarrow x<y$

12. (1) $4x-y=1 \quad \& \quad x+y=4$
    (2) $x+y=4 \rightarrow x=1$
    (3) $4x-y=1 \rightarrow y=3$
    (4) $\neg(x=1 \quad \& \quad y=3)$

13. (1) $x=2 \lor x=3$
    (2) $x \neq 2 \lor x \neq 3$

14. (1) $2x+y=4 \leftrightarrow x+2y=5$
    (2) $(2x+y=4 \rightarrow x+2y=5) \rightarrow x=1$
    (3) $x \neq 1 \lor y=2$
    (4) $y=2 \rightarrow 2x+y=4$

15. (1) $x=y \rightarrow x<z$
    (2) $x \nless z \quad \& \quad (x=y \lor y<z)$
    (3) $y<z \rightarrow x<z$

16. (1) $3+4=12 \leftrightarrow 3 \times 4 = 22$
    (2) $3+4=12 \lor 3+3=11$
    (3) $\neg(4+4=10 \lor 3 \times 4 = 22)$
    (4) $4+4=10 \rightarrow 3+3=11$

▶ 3.7 *Indirect Proof*

We will use the rule of conditional proof and the notion of contradiction to introduce a new method of proof, the *indirect proof*. This proof may also be called *proof by contradiction* or *reductio ad absurdum* proof.* The method of proof by contradiction is easily explained. By *modus tollendo tollens* we can derive the negation of the antecedent of a conditional when we know that the consequent is false. If the consequent is a contradiction we know it is logically false. So from $P \rightarrow (Q \& \neg Q)$ we can derive $\neg P$. This is the Law of Absurdity (Ab).

---

* *Reductio ad absurdum* means "to reduce to the absurd." A contradiction is often called an absurdity. Thus to prove a contradiction is to reduce a set of premises to absurdity.

The example below illustrates this rule. Suppose we want to prove ¬P.

(1)	¬Q ∨ R	P
(2)	P → ¬R	P
(3)	Q	P
(4)	P	P
(5)	¬R	PP 2, 4
(6)	¬Q	TP 1, 5
(7)	Q & ¬Q	A 3, 6
(8)	P → Q & ¬Q	CP 4, 7
(9)	¬P	Ab 8

Notice that this is a conditional proof. The premise we added is the negation of the conclusion we wanted. In every indirect proof we derive a contradiction, as in line (7) of the example, from an added premise, line (4), and we always infer the negation of the added premise, line (9). If we used only the law of absurdity then we would always need a conditional step, like line (8), before we could infer the negation of the added premise. But the rule of indirect proof allows us to combine this conditional proof step and the use of the law of absurdity into one step. We need not write the CP line.

The rule for indirect proof (RAA) is summarized:

If a contradiction can be derived from a set of premises and the negation of S, *then* S *can be derived from the set of premises alone.*

We use the letters 'RAA' (for *reductio ad absurdum*) to refer to the rule for an indirect proof.

The steps used in an indirect proof are:

(1) Introduce the negation of the desired conclusion as a new premise.
(2) From this new premise, together with the given premises, derive a contradiction.
(3) State the desired conclusion as a logical inference derived from the original premises.

The next example illustrates the use of an indirect proof to arrive at a desired conclusion. The conclusion desired is ¬D.

(1)	D → W	P
(2)	A ∨ ¬W	P
(3)	¬(D & A)	P
(4)	D	P
(5)	W	PP 1, 4
(6)	A	TP 2, 5
(7)	¬D ∨ ¬A	DL 3
(8)	¬A	TP 4, 7
(9)	A & ¬A	A 6, 8
(10) ¬D		RAA 4, 9

Let us examine the derivation above and consider the three steps always included in an indirect proof. The first step, the introduction of the negation of the desired conclusion as a new premise, is seen in line (4), for D is the negation of ¬D. The second step, the derivation of a contradiction from the new premise together with the given premises is seen in lines (5) through (9). Line (9), A & ¬A, is the derived contradiction. The third step, stating the desired conclusion as an inference from the premises, is found in line (10). Our conclusion, ¬D, is derived by RAA from the premises and that step is from lines (4) and (9), the added premise and the contradiction derived from it. With the adding of a new premise in line (4) the steps of the proof were moved to the right.

A subordinate proof indicates that any derivation from the set of premises plus the added premise depends upon the added premise in addition to the original three. The conclusion is moved back in line with the original set of premises to show that it is logically derived from the original set of premises alone. An indentation or subordinate proof can be ended only when applying CP or RAA.

Let us consider another example of the use of an indirect proof. The argument is

> If John plays first base and Bill pitches against us then State will win.
> Either State will not win or the team will end up at the bottom of the league.
> The team will not end up at the bottom of the league.
> Furthermore, John will play first base.
> Therefore, Bill will not pitch against us.

To emphasize that any contradiction you can derive is acceptable, two formal proofs by RAA are given.

(1)	J & B → S		P
(2)	¬S ∨ T		P
(3)	¬T		P
(4)	J		P
(5)		B	P
(6)		J & B	A 4, 5
(7)		¬S	TP 2, 3
(8)		¬(J & B)	TT 1, 7
(9)		(J & B) & ¬(J & B)	A 6, 8
(10)	¬B		RAA 5, 9

(1)	J & B → S		P
(2)	¬S ∨ T		P
(3)	¬T		P
(4)	J		P
(5)		B	P
(6)		¬S	TP 2, 3
(7)		¬(J & B)	TT 1, 6
(8)		¬J ∨ ¬B	DL 7
(9)		¬B	TP 8, 4
(10)		B & ¬B	A 5, 9
(11)	¬B		RAA 5, 10

Notice in the second proof that line (9) is ¬B, the desired conclusion. But the proof is not yet complete because it is in a subordinate proof, it depends on the added premise, not just the original premises.

In the example just given the conclusion was derived by indirect proofs. The addition of **B** as a premise led to a contradiction and therefore we could conclude the negation of **B**, ¬B. In the same example we might have derived ¬B by a direct proof without adding a premise. Either method is correct. As in any game, many different moves are permitted by the rules. The point is to make moves which will lead to the goal, which is the desired conclusion.

There is no general strategy or rule which can tell you exactly when to use a direct proof and when to use an indirect proof. Usually an indirect proof is suggested by a set of premises that seem to provide no starting point. In such a situation, by adding a premise, the negation of the desired conclusion, we may find a place to begin. The second example in this section on indirect proofs illustrates this dilemma. We were faced with only conditionals and disjunctions in the given premises so that we had no starting place. By adding a premise, however, we had an atomic sentence which opened the way to other moves leading eventually to the conclusion.

EXERCISE 12

A. Prove that the following conclusions are valid by using an *indirect proof*.

1. Prove: ¬P
   (1) ¬(P & Q)
   (2) P → R
   (3) Q ∨ ¬R

2. Prove: ¬T
   (1) T → ¬S
   (2) F → ¬T
   (3) S ∨ F

3. Prove: R
   (1) ¬(P & Q)
   (2) ¬R → Q
   (3) ¬P → R

4. Prove: ¬(A & D)
   (1) A → B ∨ C
   (2) B → ¬A
   (3) D → ¬C

5. Prove: ¬E ∨ M
   (1) S ∨ O
   (2) S → ¬E
   (3) O → M

6. Prove: ¬T
   (1) P ∨ Q
   (2) T → ¬P
   (3) ¬(Q ∨ R)

7. Prove: ¬(T ∨ S)
   (1) ¬R ∨ ¬B
   (2) T ∨ S → R
   (3) B ∨ ¬S
   (4) ¬T

8. Prove: ¬P
   (1) P → ¬S
   (2) S ∨ ¬R
   (3) ¬(T ∨ ¬R)

9. Prove: ¬S ∨ ¬T
   (1) ¬P → ¬S
   (2) ¬P ∨ R
   (3) R → ¬T

10. Prove: R
    (1) T & R ↔ ¬S
    (2) ¬S → T
    (3) ¬R → ¬S

11. Prove: $\neg(y=1 \rightarrow x^2 \not> xy)$
    (1) $x=1 \lor \neg(x+y=y \lor x \not> y)$
    (2) $x>y \rightarrow x^2>xy \ \& \ y=1$
    (3) $x \neq 1$

12. Prove: $\neg(x=2 \;\leftrightarrow\; x=y)$
    (1) $x<y \;\rightarrow\; xy=x$
    (2) $x\neq y \;\;\&\;\; xy\neq x$
    (3) $x\nless y \;\lor\; y=1 \;\rightarrow\; x=2$

13. Prove: $2x=12 \;\rightarrow\; y=4$
    (1) $2x+3y=24$
    (2) $(x=6 \;\rightarrow\; y=4) \;\lor\; 2x=12$
    (3) $(2x=12 \;\rightarrow\; x=6) \;\lor\; 2x+3y\neq24$
    (4) $x\neq6$

14. Prove: $x=0$
    (1) $\neg(y\neq1 \;\lor\; z\neq-1)$
    (2) $(x<y \;\;\&\;\; x>z) \;\;\&\;\; z=-1 \;\rightarrow\; x=0$
    (3) $\neg(y=1 \;\lor\; x=0) \;\lor\; (x<y \;\;\&\;\; x>z)$

15. Prove: $x=0$
    (1) $y=1 \;\rightarrow\; x=0 \;\lor\; x>y$
    (2) $z=-1 \;\rightarrow\; x=0 \;\lor\; x<z$
    (3) $x\ngtr y$
    (4) $x\nless z$
    (5) $y=1 \;\lor\; z=-1$

B. In Section **A** above you were required to derive the conclusion by using indirect proofs. Could a direct proof be used in any of the examples in Section **A**? If so, show a direct proof in the standard form for each example where the direct method is possible.

C. Each of the following derivations contains errors. Find all errors and make the necessary corrections.

1. Prove: $\neg(V \;\&\; R)$
(1)	$V \rightarrow T$	P
(2)	$T \rightarrow S$	P
(3)	$R \rightarrow \neg S$	P
(4)	$V \;\&\; R$	P
(5)	T	TP 1, 4
(6)	$\neg S$	PP 2, 5
(7)	R	S 4
(8)	$\neg\neg S$	PP 3, 7
(9)	$\neg S \;\&\; \neg\neg S$	A 6, 8
(10)	$V \;\&\; R \rightarrow \neg S \;\&\; \neg\neg S$	CP 4, 9
(11)	$V \;\&\; R$	RAA 10

2. Prove: ¬(T ∨ P)

(1)	¬T ∨ ¬R	P
(2)	¬R → S	P
(3)	¬S & ¬P	P
(4)	¬R	S 1
(5)	¬S	PP 2, 4
(6)	¬¬R	TT 2, 5
(7)	T	TP 1, 6
(8)	¬P	S 3
(9)	T & ¬P	A 7, 8
(10)	¬(T ∨ P)	DL 9

 3.8 *Summary*

We learn to prove the validity of inference in much the way we learn to play a game. We start with given premises and our goal is to reach a particular conclusion. The way in which we proceed toward the goal is to derive sentences from the sentences we already have by use of the rules of inference. Each move we make must be permitted by a rule.

In a formal proof we must justify every step we take by reference to a rule of inference. We also tell from what sentences our new sentence was derived. The rules permit us to make many moves but the strategy is to make the moves which will lead toward the goal, the desired conclusion.

The idea of inference is summed up in this way:

From premises that are true we reach only conclusions that are true.

Since our rules of valid inference permit us to derive only true conclusions from true premises, if we find a case in which a false conclusion is derived from true premises we know that the inference is not valid. Thus, if we assign truth values to the atomic sentences so that premises are true and conclusion false we have proved invalidity of the argument.

In addition to the direct method of proof, we can often use the rule which permits us to introduce a premise in our proof. In a conditional proof, for example, we introduce the antecedent of the conclusion (when it is a conditional sentence) as a premise and if we can derive the consequent then we have proved the conditional sentence follows from the original premises. In an indirect proof, if by introducing the negation of our desired conclusion we can derive a contradiction of the form **P** & **¬P**, then we can assert the desired conclusion by the *reductio ad absurdum* rule.

Sometimes we do not wish to derive a conclusion from a set of premises but we wish to determine whether a set of premises is consistent or inconsistent. We prove premises inconsistent if we can derive a contradiction of the form **P** & **¬P** from them. We then know that the premises cannot all be true at once. To prove the consistency of premises, we find a truth assignment in which all the premises *are* true at once.

The theory of inference with which we have been concerned until now is the sentential theory of inference.

EXERCISE 13

Review Exercise

A.   For each of the following, first express in logical symbols and then state and prove whether it is logically false by deriving a contradiction, or possibly true as shown by an assignment of truth values.

1. There are fifty states in the United States but if the United States had not bought Alaska from Russia then there would not be fifty states in the United States.
2. If it is not the case that either Andy plays trumpet or he does not play an instrument, then Andy plays trumpet.
3. Harry likes Latin and he doesn't like Latin.
4. $(¬P \& Q) \& (Q \rightarrow P)$.
5. $(P \& Q) \rightarrow (P \rightarrow Q \lor R)$
6. $¬[(¬P \lor Q) \lor (¬P \lor ¬Q)]$
7. $P \rightarrow ¬P$
8. $¬(P \lor ¬P)$
9. $(x=3 \quad \rightarrow \quad x<4) \quad \& \quad x \neq 3$
10. $¬((1=2 \quad \rightarrow \quad 2=1) \quad \rightarrow \quad 1 \neq 2)$

B.   Express the following sets of sentences in logical symbols and then state and prove whether they are consistent or inconsistent.

1. If Alice is well then her temperature is not 98.6.
   Alice is well if and only if her temperature is 98.6.
   But if she is well and her temperature is not 98.6 then she is not well.
2. Nitrates are formed from free atmospheric nitrogen or nitrates are formed from decaying protein in the soil.

If nitrates are formed from free atmospheric nitrogen then this process of forming nitrates is called nitrogen fixation.

The process does not include release of ammonia from decaying protein.

If the process does not include release of ammonia from decaying protein then the nitrates are not formed from decaying protein in the soil.

3. It is not the case that either Jan buys a tennis racket or she buys tennis balls.

   Either Jan buys tennis balls or she is not satisfied with the racket she already has.

   If Jan does not buy a tennis racket then she is satisfied with the racket she already has.

4. If Bill visits Bob then Paul visits Peter.

   If Paul does not visit Peter then either Bill and Paul go to the movies or they finish their English assignment.

   But Bill and Paul do not finish their English assignment.

   Furthermore, Bill and Paul go to the movies and Bill doesn't visit Bob.

5. If a wire in one electric circuit melts and the wire in another does not melt then the first wire has higher resistance or it had a higher magnitude of current flowing through it. If the first wire had a higher resistance then a greater quantity of electric energy in the first circuit was converted to heat.

   If it had a higher magnitude of current flowing through it then a greater quantity of electric energy in the first circuit was converted to heat.

   It is not the case that a greater quantity of electric energy in the first circuit was converted to heat.

   It is not the case that a wire in one electric circuit melted and the wire in another did not melt.

6. (1) $E \rightarrow G \lor H$
   (2) $J \rightarrow \neg H$
   (3) $G \,\&\, H$
   (4) $G \rightarrow \neg E$

7. (1) $P \,\&\, Q$
   (2) $P \rightarrow \neg R$
   (3) $\neg R \,\&\, S \rightarrow \neg Q$
   (4) $Q \rightarrow S$

8. (1) $U \rightarrow W \lor (R \lor S)$
   (2) $W \lor R \rightarrow \neg U$
   (3) $U \,\&\, \neg S$

9. (1) $P \rightarrow Q \lor R$
   (2) $Q \rightarrow \neg P$
   (3) $R \rightarrow \neg P$

10. (1) $x \neq y$ & $y \neq z$
    (2) $\neg(x = z \ \lor \ x < z)$
    (3) $z = 2 \ \rightarrow \ y = z$
    (4) $x < z \ \lor \ z = 2$

C. In each example below, prove in complete standard form that the conclusion follows from the given premises or give a truth assignment that shows the conclusion does not logically follow.

1. If the stick begins to beat the dog then the dog begins to bite the pig.
   If the dog begins to bite the pig then the pig will jump over the stile.
   The stick does begin to beat the dog.
   Therefore, the pig will jump over the stile.

2. If the brakes fail or the road is icy then the car will not stop.
   If the car was inspected then the brakes do not fail.
   But the car was not inspected.
   Therefore, the car will not stop.

3. If a dam is built to supply hydroelectric power then the country's manufacturing industry will increase greatly.
   If a dam is not built to supply hydroelectric power then the economy must be based almost totally on agricultural products.
   Therefore, either the country's manufacturing industry will increase greatly or the economy must be based almost totally on agricultural products.

4. Either there are not many cats or there are few mice.
   There are many flowers.
   If there are few mice then if there are many flowers there will be many bumble bees.
   Therefore, there are many cats or there will be many bumble bees.

5. If the point on the line represents an integer then the number can be described by an infinite decimal or by a pair of infinite decimals.
   Either the number can be described by a finite decimal or the number can be described by either an infinite decimal or by a pair of infinite decimals.
   The number cannot be described by a finite decimal.
   Therefore, the point on the line represents an integer.

6. A dense gas (hydrogen chloride) is placed in a bottle and over this is placed a bottle containing a low density gas (ammonia). If the gases mix by diffusion then the hydrogen chloride has risen and the ammonia has descended.

If the hydrogen chloride has risen and the ammonia has descended then the movement of the gases is opposite to that caused by gravity.

If the movement of the gases is opposite to that caused by gravity then the movement must be due to molecular motion. Therefore, the movement must be due to molecular motion.

7. Prove: $\neg B \rightarrow \neg Q$
   (1) $R \rightarrow N$
   (2) $K \rightarrow B \lor R$
   (3) $Q \lor M \rightarrow K$
   (4) $\neg N$

8. Prove: $\neg J \lor C$
   (1) $J \lor S \rightarrow C \& V$

9. Prove: P
   (1) $R \lor Q \rightarrow \neg P$
   (2) $S \rightarrow \neg Q$
   (3) $\neg R \& S$

10. Prove: $B \lor C$
    (1) $A \rightarrow B$
    (2) $C \rightarrow D$
    (3) $A \lor D$

11. Prove: $G \lor J \rightarrow H \lor K$
    (1) $G \rightarrow H$
    (2) $J \rightarrow K$

12. Prove: C
    (1) $W \rightarrow F$
    (2) $F \& C \leftrightarrow W$
    (3) $\neg C \rightarrow W$

13. Prove: $C \& \neg D$
    (1) $A \& C \rightarrow B$
    (2) $\neg A \lor (C \lor D)$
    (3) $A \& B$

14. Prove: $P \& Q$
    (1) $P \leftrightarrow Q$
    (2) $P \lor Q$

15. Prove: $\neg S \rightarrow \neg R$
    (1) $R \rightarrow \neg Q$
    (2) $R \lor Q$
    (3) $R \rightarrow S$

16. Prove: $M \leftrightarrow N$
    (1) $M \lor N$
    (2) $N \leftrightarrow (M \rightarrow P)$
    (3) $P \lor (N \& Q)$
    (4) $Q \leftrightarrow (P \rightarrow N)$

17. Prove: $x = 3$
    (1) $x^2 - 5x + 6 = 0 \lor x^2 - 7x + 12 = 0$
    (2) $x^2 - 7x + 12 = 0 \leftrightarrow x = 3 \lor x = 4$
    (3) $x^2 - 5x + 6 = 0 \leftrightarrow x = 3 \lor x = 2$

18. Prove: $z=3$
    (1) $x<y$ & $y<z$ → $x<z$
    (2) $(y<z$ → $x<z)$ → $z=3$
    (3) $x<y$

19. Prove: $x \not< z$
    (1) $x=y$ → $x\neq0$
    (2) $x=y$ & $z=-1$
    (3) $x<z$ → $x\neq0$

20. Prove: $\neg(2x+2y=8$ ↔ $y=2)$
    (1) $y=2$ → $4x+y=6$
    (2) $y=3$ → $2x+2y=8$
    (3) $\neg(4x+y=6$ ∨ $y\neq3)$

R E V I E W   T E S T

I. Prove, by truth assignment or derivation, whether each of the formulas is possibly true or logically false.

    a. (P ↔ Q) & (P & ¬Q)
    b. P & ¬[(P ∨ Q) ∨ R]
    c. ((P → Q) → Q) → Q
    d. $(x \not< y$ → $y=x)$ & $(y=x$ → $x<y)$
    e. $x=3$ & $\neg(x\neq y$ ∨ $x=3)$

II. Prove by truth assignment or derivation whether each of the following sets of premises is consistent or inconsistent.

a. (1) ¬P ∨ Q
    (2) ¬R → ¬Q
    (3) P → R

b. (1) P → (Q → R)
    (2) P & ¬R
    (3) Q

c. (1) P → Q
    (2) R → S
    (3) P ∨ G
    (4) Q & S

d. (1) A → (B & C)
    (2) D → (A ∨ E)
    (3) E → (C → F)
    (4) ¬D ∨ ¬F

e. (1) P → Q ∨ ¬R
    (2) P → R
    (3) P → ¬Q
    (4) P

f. (1) $x\neq1$
    (2) $x+y\neq y$ → $x>0$
    (3) $\neg(x>0$ ∨ $y\neq1)$
    (4) $x+y=y$ ↔ $x\neq1$

III.   Derive, if possible, the desired conclusion from the given premises in each of the following. If the conclusion does not follow write 'invalid' and give a truth assignment that proves it.

a. Either France was a monarchy in 1780 or France was a republic in 1780.

If France was a monarchy in 1780, then the American Revolution preceded the French Revolution.

France was not a republic in 1780.

Therefore, the American Revolution preceded the French Revolution.

b. If angle alpha equals angle beta, then angle beta equals 45 degrees.

If angle beta equals 45 degrees then angle theta equals 90 degrees.

Either angle beta is a right angle or angle beta does not equal 90 degrees.

Angle theta is not a right angle.

Therefore, angle alpha does not equal angle beta.

c. If Bob selects Tom as a campaign manager then he will win the election. If Bob does not win the election then he will continue as editor of the newspaper.

If he continues as editor of the newspaper then Jim will be associate editor.

Therefore, either Jim will be associate editor or Bob does not select Tom as his campaign manager.

d. Prove: $\neg P$
 (1) $P \rightarrow Q$
 (2) $R \vee \neg Q$
 (3) $\neg P \vee \neg R$

e. Prove: $C \vee \neg A$
 (1) $A \rightarrow B$
 (2) $\neg B \rightarrow C \vee D$
 (3) $\neg D$

f. Prove: $\neg E$
 (1) $E \rightarrow (G \vee H)$
 (2) $G \rightarrow (H \rightarrow K)$
 (3) $\neg L$

g. Prove: N
 (1) $L \vee (M \& N)$
 (2) $L \rightarrow N$

h. Prove: $A \rightarrow (C \rightarrow E)$
 (1) $A \vee B \rightarrow (C \vee D \rightarrow E)$

i. Prove: $P \rightarrow S$
 (1) $P \rightarrow Q$
 (2) $P \rightarrow (Q \rightarrow R)$
 (3) $Q \rightarrow (R \rightarrow S)$

j. Prove: $x > y \rightarrow x = 5$
   (1) $x - y = 2 \rightarrow (y = 3 \rightarrow x = 5)$
   (2) $x \not> y \lor x - y = 2$
   (3) $y = 3$

k. Prove: $xy = y \rightarrow y \neq 2$
   (1) $x \not> y \rightarrow y = 2$
   (2) $xy = y \lor x^2 = y \rightarrow x = 1$
   (3) $x^2 = y \rightarrow y \neq 2$
   (4) $\neg(x > y \lor x \neq 1)$

l. Prove: $x = 1$
   (1) $x = 1 \leftrightarrow x = y$
   (2) $x = y \lor y \neq 1$
   (3) $y \neq 1 \ \& \ x < y$

IV. Complete the following formal proofs by filling in for each line the abbreviation of the rules used and the numbers of any lines to which the rules are applied.

Example: (1) $\neg R \rightarrow S$ P
          (2) $\neg R$ P
          (3) S PP 1, 2

a. (1) $x < y \ \& \ y \neq z$
   (2) $y \neq z$

b. (1) $S \leftrightarrow \neg R$
   (2) $(S \rightarrow R) \ \& \ (R \rightarrow S)$

c. (1) $\neg R \lor \neg R$
   (2) $\neg R$

d. (1) $\neg S \lor P$
   (2) $\neg P$
   (3) $\neg S$
   (4) $\neg P \rightarrow \neg S$

e. (1) $\neg A \rightarrow \neg(B \lor \neg C)$
   (2) $\neg\neg B$
   (3) B
   (4) $B \lor \neg C$
   (5) $\neg\neg A$

f. (1) $M \rightarrow \neg(P \lor Q)$
    (2) $\neg(P \lor Q) \rightarrow N$
    (3) $M \rightarrow N$

g. (1) $x+y=3 \quad \lor \quad (y=2 \quad \rightarrow \quad x+y=5)$
    (2) $(y=2 \quad \rightarrow \quad x+y=5) \quad \lor \quad x+y=3$

h. (1) $R \rightarrow \neg Q$
    (2) $Q \lor P$
    (3) $\neg(\neg R \lor P)$
    (4) $R \ \& \ \neg P$
    (5) R
    (6) $\neg Q$
    (7) P
    (8) $\neg P$
    (9) $P \ \& \ \neg P$
   (10) $\neg R \lor P$

i. (1) $(x \neq 3 \quad \lor \quad y=2) \quad \& \quad x>y$
    (2) $(x=3 \quad \rightarrow \quad x=y) \quad \rightarrow \quad x \not> y$
    (3) $y=2 \quad \rightarrow \quad x=y$
    (4) $x \neq 3 \quad \lor \quad y=2$
    (5) $x=3$
    (6) $y=2$
    (7) $x=3 \quad \rightarrow \quad y=2$
    (8) $x=3 \quad \rightarrow \quad x=y$
    (9) $x \not> y$
   (10) $x>y$
   (11) $x>y \quad \& \quad x \not> y$
   (12) $\neg(y=2 \quad \rightarrow \quad x=y)$

j. (1) $\neg R \lor S$
    (2) $\neg R \rightarrow A \ \& \ B$
    (3) $S \rightarrow C$
    (4) $(A \ \& \ B) \lor C$

CHAPTER FOUR
TRUTH TABLES

▶ 4.1 *Truth Tables*

A method often more convenient than the diagram for analyzing the truth values of sentences is that of putting all the possibilities for truth or falsity into the form of a table. In fact, all of our truth-functional rules of usage for molecular sentences may be summarized in table form. These *basic truth tables* tell us at a glance whether a molecular sentence is true or false if we know the truth or falsity of the sentences that make it up. The basic truth tables for all five sentential connectives are shown below. If we know the truth values of sentence **P** and sentence **Q** we find the line which shows that particular combination of truth values and in the same line under the molecular sentence we will find the truth value for it.

Negation			Conjunction				Disjunction		
P	**¬P**		**P**	**Q**	**P & Q**		**P**	**Q**	**P ∨ Q**
T	F		T	T	T		T	T	T
F	T		T	F	F		T	F	T
			F	T	F		F	T	T
			F	F	F		F	F	F

Conditional				Equivalence		
P	**Q**	**P → Q**		**P**	**Q**	**P ↔ Q**
T	T	T		T	T	T
T	F	F		T	F	F
F	T	T		F	T	F
F	F	T		F	F	T

All of the rules of usage we have learned are summarized in the tables above. If you are uncertain about any of these rules you can use these tables as reference tables.

In Chapter 3 we said that we needed a general method of determining validity so that we could be certain of the validity of any suggested rule of inference. The truth tables provide us with a mechanical method for testing validity. We can check the validity of any inference without reference to a particular named rule which permits that inference.

Before developing this check, let us return briefly to the notion of a valid inference itself. If an inference is valid, then in every possible interpretation or truth assignment if the premises are true the conclusion of the argument will also be true. The truth table gives every possible truth assignment. First, list all of the possible combinations of truth values for the atomic sentences included in the example. Second, determine the truth values for all of the *premises* and the conclusion of the argument. Third, find any lines that show all premises as true sentences; if the conclusion is also true for *every* such line then the argument is valid. But if there is *any* line for which all of the premises are true and the conclusion is false, the argument is invalid and the conclusion does not follow logically.

Let us consider an example of a rule of inference which we already know. We shall use a truth table to check the validity of the rule *modus tollendo ponens*. The premises are of the form $P \lor Q$ and $\neg P$. Therefore, we need to find all possible truth values of those two sentences. To do this, we first put down all the possible combinations of truth or falsity for the atomic sentences that make up those molecular sentences. The atomic sentences are sentence P and sentence Q. The number of possible combinations of truth or falsity depends upon the number of atomic sentences involved. In this case we have two atomic sentences and since for each of them there are two possible truth values, the number of lines in the truth tables will be 2×2 or 2^2.* We construct the truth table in the following way:

P	Q	P ∨ Q	¬P
T	T	T	F
T	F	T	F
F	T	T	T
F	F	F	T

---

* If there are three atomic sentences then there are twice as many or eight possible combinations of truth or falsity. Since there are two possible truth values for each atomic sentence then for three atomic sentences we have $2 \times 2 \times 2$ or 2^3 combinations. The general rule is that if there are n atomic sentences then there are 2^n combinations of possible truth values.

The method for filling in the truth table above is the following:
We begin by filling in all the combinations of truth or falsity under
sentences P and Q. The truth value of molecular sentences depends
upon the truth value of sentences P and Q. Therefore, in filling in the
column under each molecular sentence we refer to the truth values of
its parts. For example, in the first line we have P as true and Q as true.
Therefore, P ∨ Q is a true sentence; and since P is a true sentence ¬P
is false. On the other hand, in the last line of the table P and Q are both
false sentences so the sentence P ∨ Q must be false, and ¬P is true.

The next step is to look for lines in which *all* the premises of the
argument are true. In this case, the premises of the argument are P ∨ Q
and ¬P. Looking at the truth table we see that the premises make up
the last two columns of the table. Looking down the columns under
those headings we find only one case where both of the premises are
true together. This is in the third line. To show that on this line all the
premises are true together, we circle the T's for the premises in the
third line. Since a valid inference requires that in all cases where
premises are true the conclusion will also be true, the conclusion should
be true in the third line too if the argument is valid. The conclusion
of the argument is Q. We now check the column under Q for the truth
value of the third line. Since we find that it is true, then we know that
the inference in question is a valid one. To make this easy to see in the
table we put a square around the truth assignment of the conclusion
on each line where all the premises are true; that is, each line where
the premise truth values are circled.

To provide a contrast with the valid inference *modus tollendo ponens*
we next consider the fallacy of affirming the consequent, discussed in
the last chapter. Our aim is to show how truth-table analysis may be
used to demonstrate that it is a fallacy. This fallacious inference has
the form

$$P \rightarrow Q$$
$$Q$$
$$\overline{}$$
$$P$$

The premises are P → Q and Q; the conclusion is P. Because we have
two atomic sentences, P and Q, the truth table must have four lines.
Both of the premises are true in lines (1) and (3) of the table, but only
in line (1) is the conclusion P also true. Because of the F under P in
the third line we know from the table that the inference is fallacious.

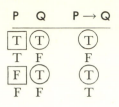

P   Q   P → Q

It is important to understand exactly why the truth table shows that this inference is fallacious. Looking at line (3) we note that **P** is false and **Q** is true. If we choose any two atomic sentences with these respective truth values, we may construct the true premises **P → Q** and **Q** and the false conclusion **P**. In this case it is at once apparent that the conclusion is false for it is just the atomic sentence **P**. For example, let **P** = '1 = 2' and **Q** = '0 = 0'. Then the fallacious inference would be

> If 1 = 2 then 0 = 0
>
> 0 = 0
>
> Therefore 1 = 2.

In Chapter Three, we suggested an example of valid inference which had not been introduced as a rule of inference. It was suggested that from sentence **P → Q** we could infer sentence **¬P ∨ Q**. We can check the validity of that inference by constructing the appropriate truth table. The two atomic sentences are sentence **P** and sentence **Q**; so we will begin by filling column **P** and column **Q**. Next we obtain the truth values for the premise, **P → Q**. In order to get the truth values for the disjunction **¬P ∨ Q**, which is our conclusion, we must first find the truth values for **¬P**.

P	Q	'¬P	P → Q	¬P ∨ Q
T	T	F	T	T
T	F	F	F	F
F	T	T	T	T
F	F	T	T	T

The procedure for filling in the columns in the truth table above is: Fill in the truth values for column 1 and column 2. We get column 3, **¬P**, by referring to the truth values of column 1. We get the values for column 4 by looking at the values of column 1 and column 2 together.

Finally, we get the values of column 5 by considering column 2 and column 3 together.

Column 4 represents the only premise in the example of inference. We look for the cases in which that premise is true. For this premise we have *true* as a truth value in lines (1), (3), and (4). Therefore, we circle the three T's. If the inference is a valid one then the conclusion will be true in each of those lines. Checking column 5, which is our conclusion, we find the letter 'T' in every line in which we found the letter 'T' for the premise, as shown by the T's in squares. Thus we conclude that the suggested inference is a valid one.

To show the power of this method of analysis it will be useful to consider a more complicated example about which we may not be as certain of validity or invalidity in advance. Consider the following mathematical argument

$$\text{If } x = 0 \text{ and } y = z \text{ then } y > 1$$
$$y \not> 1$$
$$\text{Therefore } y \neq z.$$

We want to know if this argument is valid. Three atomic sentences occur in the argument, which we shall symbolize as

$$A = {}'x = 0{}'$$
$$B = {}'y = z{}'$$
$$C = {}'y > 1{}'.$$

Because each atomic sentence may be true or false, there are $2^3 = 8$ truth combinations and thus eight lines in the truth table.*

In terms of the symbols A, B, and C the argument we are considering may be symbolized

$$A \ \& \ B \rightarrow C$$
$$\neg C$$
$$\overline{}$$
$$\neg B$$

---

* To be certain of obtaining all eight combinations and not writing some twice, the following systematic procedure may be useful. Let the first three columns be headed A, B, and C. Under C alternate writing T and then F. Under B, alternate writing two T's then two F's. And finally under A alternate writing four T's then four F's. This manner of assigning truth values to the atomic sentences will give all possible combinations without any duplicate rows.

The truth table for this is

A	B	C	A & B	A & B → C	¬C	¬B
T	T	T	T	T	F	F
T	T	F	T	F	T	F
T	F	T	F	T	F	T
T	F	F	F	(T)	(T)	[T]
F	T	T	F	T	F	F
F	T	F	F	(T)	(T)	[F]
F	F	T	F	T	F	T
F	F	F	F	(T)	(T)	[T]

As the circles show, for three of the eight lines both premises are true, but as the squares show, for one of these three lines the conclusion is not true. Line (6) shows that the argument is fallacious. If we take x to be 1, and y and z to be 0, this may easily be seen:

$$\text{If } 1=0 \text{ and } 0=0 \text{ then } 0>1$$
$$0 \not> 1$$
$$\text{Therefore, } 0 \neq 0$$

The first premise is true because the antecedent is false, and the second premise is clearly true, but the conclusion is just as clearly false.

In examining this truth table note that column 4 represents an intermediate step. A & B is neither an atomic sentence, a premise or a conclusion. It is a molecular sentence that is part of one of the premises. It illustrates a rule that should always be followed: The truth table analyzing an argument should have a column (a) for each atomic sentence, and (b) for each molecular sentence occurring in the argument. Requirement (b) means there will be a column for each occurrence of a connective in each sentence except that we do not need to repeat if the same connective combines the same sentences twice. So if we have A & B and A & B → C we would need only one A & B column. In the present case this means seven columns, three for the atomic sentences A, B, and C, and four for the molecular sentences A & B, A & B → C, ¬C and ¬B.

EXERCISE 1

A.   Show by a truth table which of the following examples of inference is valid. Exhibit the entire truth table and write the words 'valid' or 'invalid' beside each.

1. If Becky is late then Christine is early.
    If Becky is not late then Christine is not early.
    Therefore, either Becky is late or Christine is early.

2. If I am eighteen then I am older than Paul.
    If I am not eighteen then I am younger than George.
    Therefore, either I am eighteen or I am younger than George.

3. Either Johnson does not deliver the merchandise or the contract is considered legal.
    Therefore, if Johnson does deliver the merchandise then the contract is considered legal.

4. If I were the President then I would live in Washington, D.C.
    I am not the President.
    Therefore, I do not live in Washington, D.C.

5. A hydrogen atom has one proton in its nucleus and the atomic number of hydrogen is 1.
    Therefore, a hydrogen atom has one proton in each nucleus if and only if the atomic number of hydrogen is 1.

6. Continually planted soil becomes depleted if and only if measures are not taken to restore the minerals extracted by the crops.
    Therefore, either continually planted soil becomes depleted or measures are taken to restore the minerals extracted by the crops.

7. Either AB is greater than BC or AB is not equal to CD.
    Therefore, AB is not equal to CD and AB is greater than BC.

8. $(P \lor Q)$ & $\neg Q$.
    Therefore, P.

9. $\neg Q \rightarrow \neg P$.
    Therefore, $P \rightarrow Q$.

10. $P \rightarrow \neg Q$.
    $\neg Q$.
    Therefore, $\neg P$.

B.   Complete the truth table below to show that the Law of Hypothetical Syllogism is a good rule.

P	Q	R	P → Q	Q → R	P → R
T	T	T			
T	T	F			
T	F	T			
T	F	F			
F	T	T			
F	T	F			
F	F	T			
F	F	F			

C.   Show a truth table to prove that $\neg P$ & $\neg Q$ follows logically from $\neg(P \lor Q)$.

D.   Show a truth table to prove that the Rule of Adjunction is a good rule of inference.

E.   Show a truth table to prove that the rule, *modus tollendo ponens*, is a good rule.

F.   Show by a truth table which of the following mathematical arguments are valid and which are invalid.

1. $x = 3$. Therefore, $y = 0 \rightarrow x = 3$.
2. $x \neq y \rightarrow x = y$. $y = 1 \lor x \neq y$. Therefore, $y = 1$.
3. $x < 5 \rightarrow x \neq y$. $x \neq y$ & $x < 5$. Therefore, $x \not< 5$ & $x = y$.
4. $x < 3 \rightarrow x \not< 3$. Therefore, $x \not< 3$.
5. $x = y \rightarrow x \neq y$ & $y = 2$. Therefore, $x \neq y$.
6. $x = y \rightarrow x = y$ & $y = 2$. Therefore, $x = y$.
7. $x < z \rightarrow x \neq y$. $\neg(x \not< z$ & $x = y)$. Therefore, $x < z \lor x = y$.
8. $3 \not< y \rightarrow x > y$. $x > y \rightarrow (x \not> y$ & $3 \not< y)$. Therefore, $3 < y \lor x > y$.
9. $x^2 = 4 \rightarrow x = 2$. $\neg(x = 2 \lor x^2 \neq 4)$. Therefore, $x = 2$ & $x \neq 2$.
10. $x = 2 \lor x < 2$. $x = 3 \rightarrow x \neq 2$. $x = 3 \rightarrow x \not< 2$. Therefore, $x \neq 3$.
11. $x = y \leftrightarrow y \neq 1$. $\neg(x = y$ & $y \neq 1)$. Therefore, $y \neq 1$.
12. $x \neq y \rightarrow x < 5$. $x \not< 5 \lor y < 6$. $x = y$ & $x \not< 5$. Therefore, $x = y \rightarrow y < 6$.

▶ 4.2 *Tautologies*

A molecular sentence is a *tautology* if it is true no matter what the truth values of the atomic sentences which make it up. In a tautology we can substitute for its atomic sentences any other atomic sentences, whether true or false, and the sentence itself is still true. For instance, for any atomic sentence **P**

$$P \lor \neg P$$

is a tautology. If **P** is true then **P** ∨ ¬**P** is true. Furthermore, if **P** is false, then **P** ∨ ¬**P** is still true.

We can show this by a truth table.

P	¬P	P ∨ ¬P
T	F	T
F	T	T

In a truth table, if a sentence is a tautology then every line in its column will have a 'T' as its entry. This means that the sentence is always true regardless of the combinations of truth values of its atomic sentences. We must remember that in any particular case an atomic sentence has the same truth values each time it occurs within a molecular sentence. If **P** occurs more than once in a particular sentence then it cannot be true in one case and false in another. If it is false, then it is false in all occurrences and if it is true, then it is true in all occurrences. A line of a truth table is one particular case.

Is the sentence **P** ∨ **Q** → **P** a tautology? To answer this question we may construct a truth table.

P	Q	P ∨ Q	P ∨ Q → P
T	T	T	T
T	F	T	T
F	T	T	F
F	F	F	T

In this table we obtained the third column from the first two by referring to the rules of usage for disjunctions. We obtained the last column from the first and the third by referring to the rule of usage for conditionals. If the sentence suggested is a tautology it must have

'T' as its entry in every row of the fourth column. The letter 'F' for false in the third line shows us that P ∨ Q → P is not a tautology, for it shows us that with the combination of P being false and Q being true then the sentence P ∨ Q → P is a false one. The letter 'F' in a single row of the column under the sentence in question is enough to demonstrate that the sentence is not a tautology.

A formal definition of a tautology is

> *A sentence is a tautology if and only if it remains true under all combinations of truth assignments for each of its distinct atomic sentences.*

The truth table method for determining whether a formula is a tautology uses this definition. No matter what atomic sentences we substitute for the atomic sentences in a tautology the resulting sentence will always be true. Thus we find 'T' for true as the entry in every line in the final column of the table, as in the following example.

P	Q	P & Q	¬(P & Q)	P ∨ ¬(P & Q)
T	T	T	F	T
T	F	F	T	T
F	T	F	T	T
F	F	F	T	T

EXERCISE 2

A. If P and Q are distinct atomic sentences, which of the following are tautologies? Use truth tables.

1. P ↔ Q
2. P ↔ P ∨ P
3. P ∨ Q ↔ Q ∨ P
4. (P → Q) ↔ (Q → P)
5. (P ↔ P) → P

6. P ∨ Q → P
7. P & Q → P ∨ Q
8. ¬P ∨ ¬Q → (P → Q)
9. P ∨ ¬Q → (P → ¬Q)
10. ¬P ∨ Q → (P → Q)

B. Let P, Q, and R be distinct atomic sentences. Decide by truth tables which of the following sentences are tautologies.

1. P ∨ Q
2. P ∨ ¬P
3. P ∨ Q → Q ∨ P
4. P → (P ∨ Q) ∨ R
5. P → (¬P → Q)

6. (P → Q) → (Q → P)
7. [(P → Q) ↔ Q] → P
8. P → [Q → (Q → P)]
9. P & Q → P ∨ R
10. P & Q → (P ↔ Q ∨ R)

▶ 4.3 *Tautological Implication and Tautological Equivalence*

A sentence **P** is said to *tautologically imply* a sentence **Q** if and only if the conditional **P → Q** is a tautology. Thus a *tautological implication is a tautology with the form of a conditional sentence.* The sentence 'Smith plays for Army and Johnson plays for Navy' tautologically implies 'Smith plays for Army'. This is because for any sentences **P** and **Q**, **P & Q → P** is a tautology.

The notion of tautological implication is important in our study of the validity of inference because every example of sentential inference can be expressed as a tautological implication. If we take the examples of inference found throughout Chapter 2 we can construct from each of them a conditional whose antecedent is the conjunction of premises and whose consequent is the conclusion. If the conjunction of the premises is true then the conclusion must be true. This means that to every sentential argument there corresponds a conditional and to every conditional there corresponds an argument. *The argument is valid if and only if the corresponding conditional is a tautology.*

To construct the conditional that corresponds to an argument, simply connect all the premises with &'s to make the conjunction of the premises the antecedent, then put the conclusion of the argument as the consequent. Study this example.

Argument:

> Prove: **R & S**
> (1) **P**
> (2) **P → Q**
> (3) **¬Q ∨ (R & S)**

Corresponding conditional*

$$\textbf{P \& (P → Q) \& (¬Q ∨ (R \& S)) → R \& S}$$

On the other hand, if we wish to construct the argument from the conditional, we write the consequent as the conclusion and we write the several sentences whose conjunction makes up the antecedent as the premises of the argument.

---

* Anything in the form **A & B & C** is not strictly correct because the same occurrence of **B** cannot be the right member of one conjunction and the left member of another. To be perfectly precise we would need to write (**A & B**) **& C** or **A &** (**B & C**). But since these are logically equivalent, we may omit the parentheses.

One way of talking about an argument is to say that the premises *imply* the conclusion. When we say that it is a *tautological implication* we are indicating that the corresponding conditional is a tautology and so the argument is valid. Because of this correspondence between the argument and its conditional, conditional sentences are often referred to as implications. This is particularly convenient when discussing tautological implications.

EXERCISE 3

A.   Construct the conditional corresponding to each of the following arguments.

1. Prove: R
    (1) ¬Q
    (2) ¬R → Q

2. Prove: ¬(P & ¬Q)
    (1) P → Q

3. Prove: $x=y$ → $x<z$
    (1) $x=y$ → $x=5$
    (2) $x=5$ → $x<z$

4. Prove: ¬(A ∨ B)
    (1) C & ¬D
    (2) C → ¬A
    (3) D ∨ ¬B

B.   Construct the argument (premises and conclusion) corresponding to each of the following conditionals.

1. P & (Q ∨ ¬P) → Q
2. ¬($x<0$ & $y\neq x$)   → $x\not<0$ ∨ $y=x$
3. (Q → T ∨ R) & ¬S & (R ∨ T → S) → (S → Q & ¬T)
4. (P → Q) & (P & ¬Q) → S

The rule of inference, *modus ponendo ponens*, shown as a tautological implication, would look like this:

$$(P → Q) \ \& \ P → Q.$$

The antecedent of the conditional is the conjunction of both premises. The consequent is the conclusion.

Another tautological implication is

$$(P → Q) \ \& \ ¬Q → ¬P.$$

You may recognize this tautological implication as expressing the inference we know as *modus tollendo tollens*. The antecedent is the conjunction of two premises **P** → **Q** and ¬**Q**, and the consequent is the conclusion ¬**P**.

We can prove whether any example of sentential inference is valid by expressing the inference as a conditional and determining by means of a truth table whether or not that conditional is tautological. If the implication is tautological then the inference is valid. If the implication is not tautological then the inference is invalid. Remember that the implication we are writing has as its antecedent the conjunction of all the premises and as its consequent the conclusion. An example of the truth-table test for validity where the inference is that of *modus tollendo tollens* is

P	Q	¬P	¬Q	P → Q	(P → Q) & ¬Q	(P → Q) & ¬Q → ¬P
T	T	F	F	T	F	T
T	F	F	T	F	F	T
F	T	T	F	T	F	T
F	F	T	T	T	T	T

The implication that is shown in the final column is a tautological implication because every entry in that column is seen to be T. If the implication is a tautology then the inference which derives ¬**P** from the premises **P** → **Q** and ¬**Q** is a valid inference.

Two sentences are said to be *logically equivalent* if under each possible truth assignment the two of them have the same truth value. This can be shown by a truth table. Consider **P** and ¬¬**P**

P	¬P	¬¬P
T	F	T
F	T	F

This table shows that in any one line **P** and ¬¬**P** are both true or both false. In no case is one true and the other false. So they are logically equivalent. We next use a truth table to check the equivalence of **A** & ¬**B** and ¬(¬**A** ∨ **B**)

A	B	¬A	¬B	¬A ∨ B	¬(¬A ∨ B)	A & ¬B
T	T	F	F	T	F	F
T	F	F	T	F	T	T
F	T	T	F	T	F	F
F	F	T	T	T	F	F

Checking the last two columns line by line we see that under the same truth assignments (on any one line) they have the same truth value, both true or both false. Thus we know that they are logically equivalent.

EXERCISE 4

Use truth tables to determine for each of the following pairs of sentences whether they are logically equivalent.

1. P ∨ ¬Q
   Q → P

2. $x = 1$ ∨ $x \nless 3$
   ¬($x < 3$ & $x = 1$)

3. P ∨ ¬Q → ¬P
   P → ¬P & Q

4. ¬(A → ¬B) → C
   B & ¬C → ¬A

Examine the truth table of the biconditional P ↔ Q

	P	Q	P ↔ Q
(1)	T	T	T
(2)	T	F	F
(3)	F	T	F
(4)	F	F	T

Notice in line (1) and line (4) in which **P** and **Q** have the same truth assignments (both true or both false) that the biconditional is true; and that in line (2) and line (3) in which one is true and the other false that the biconditional is false. This means that the biconditional amounts to a statement that **P** is equivalent to **Q** because P ↔ Q is true whenever they have the same truth values and false whenever their truth values are opposite. So for every pair of sentences a corresponding biconditional can be constructed by placing ↔ between them. This biconditional is true whenever the sentences have the same truth value and false when they do not. For this reason a biconditional is often called an *equivalence*.

EXERCISE 5

Construct the biconditional corresponding to each of the pairs of sentences given in Exercise 4.

If two sentences are logically equivalent, the truth value of one is always the same as the truth value of the other. This means that their corresponding biconditional will *always* be true. It is a tautology. For this reason, logically equivalent sentences are also called tautologically equivalent sentences and the corresponding biconditional (or equivalence) is called a tautological equivalence.

This gives us a second method of determining whether two sentences are equivalent. It is this: construct the corresponding biconditional (equivalence) and use a truth table to determine whether the resulting equivalence is a tautology (tautological equivalence).

EXERCISE 6

Make out the truth table for each of the biconditionals you constructed in Exercise 5 and from this tell whether each of the pairs of sentences in Exercise 4 is tautologically equivalent.

We do not ordinarily use truth tables to check the validity of an argument; usually we use the method of derivation which we learned in Chapter 2. There are two very good reasons for developing the theory of sentential inference and learning the methods of derivation of conclusion from premises. First, except for the most simple inferences, a truth table is massive and unwieldy. Even the truth tables that contain three atomic sentences become clumsy because they have eight possible combinations of truth or falsity with which to work. Most of the derivations that we can now do very quickly contain more than three atomic sentences. If a set of premises and the desired conclusion contain five distinct atomic sentences then the appropriate truth table must have 32 lines ($2^5 = 32$). It would not only have many lines but it would have very many columns because in most examples of inference there are several premises and a number of molecular sentences. Some of the examples of inference for which you can derive a step-by-step conclusion in 8 or 9 lines would take considerably more time to prove by means of the truth table. It is not only tedious to construct such a massive truth table but also difficult to avoid making a mistake. Second, the truth-table method is adequate only for proving the validity of a certain portion of logical arguments. We will need the sentential theory of inference developed in Chapter 2 in order to go on in Chapter 5 to other kinds of logical inference. The truth-table approach is not adequate to the kinds of logically valid arguments we shall examine in the following chapters. However, truth tables are a general method which can always be used for checking validity or invalidity of any sentential inference.

▶ 4.4 *Summary*

A truth table extends a truth diagram so that rather than showing the truth value of a formula for only one combination of truth assignments of its atomic sentences, it shows its truth value for every possible combination. And it can do this simultaneously for several different formulas.

To construct a truth table giving every possible combination of truth assignment to n distinct atomic letters, 2^n lines are needed. A column is needed for each distinct atomic letter and a column is needed for each occurrence of each connective that appears.

A tautology is a molecular sentence whose column in a truth table contains *no* F's.

There are two methods of using a truth table to determine whether an argument is valid. The first is to construct a truth table with a column for each premise and the conclusion and then to check line by line to see whether the conclusion is true for every line in which *all* of the premises are true. The other method is to construct the corresponding conditional and then to use a truth table to determine whether that conditional is a tautology (tautological implication).

There are two methods of using a truth table to determine whether two sentences are logically equivalent. The first is to construct a truth table with a column for each of the sentences and then to examine them line by line to see whether they always have the same truth value on each line. The other method is to construct the corresponding biconditional and then to use a truth table to determine whether that biconditional is a tautology (tautological equivalence).

EXERCISE 7

Review Exercise

A. You already know the truth values of the atomic sentences in the following molecular sentences. First translate them completely into logical symbols and then with those truth values use truth diagrams to find the truth value of each molecular sentence. (For truth diagrams and truth tables all atomic sentences, including mathematical sentences, should be represented by atomic letters.)

1. If two plus two equals five then both Columbus did not discover America and this is a logic exercise.
2. If one is not two then if two times three is not six then both nine minus five is not two and one is less than two.
3. If not either the moon is made of green cheese or cows do not have four legs then thin china dishes are easily broken if and only if spoons are used for eating food.

B.   Use the truth table of the premises and conclusion of each of the following arguments to tell whether it is valid or invalid using the first method of checking line by line.

1. $(A \rightarrow B)$ & $(A \rightarrow C)$
   $\neg A$
   Therefore: $\neg B \lor \neg C$

2. $(A \rightarrow B)$ & $(A \rightarrow C)$
   $\neg B \lor \neg C$
   Therefore: $\neg A$

C.   Determine whether the following arguments are valid, by constructing the corresponding conditional and determining by truth tables whether it is a tautological implication.

1. $x \neq y \rightarrow x = y$
   Therefore: $x = y$

2. $A \rightarrow B$ & A
   $B \lor \neg A \rightarrow C$
   Therefore: $A \rightarrow C$

D.   Let A, B, and C be any three distinct atomic sentences. Decide by truth tables which of the following are tautologies.

1. $\neg(C \& \neg(D \lor C))$
2. $(P \rightarrow Q) \rightarrow P$
3. $A \& B \rightarrow (A \leftrightarrow B \lor C)$
4. $x = 3$ & $(x \neq y \rightarrow x \neq 3)$

E.   Using the sentence P & Q as premise, determine by truth tables which of the following it tautologically implies.

1. P (Show a truth table for P & Q $\rightarrow$ P, for example)
2. P & $\neg Q$

  3. ¬P ∨ Q
  4. ¬Q → P
  5. P ↔ Q

F.   Using the sentence ¬P ∨ Q as premise, determine by truth tables which of the following it tautologically implies.

  1. P (For example, show a truth table for ¬P ∨ Q → P)
  2. Q → P
  3. P → Q
  4. ¬Q → ¬P
  5. ¬P & Q

G.   The sentence P is tautologically equivalent to which of the following?

  1. P ∨ Q
  2. P ∨ ¬P
  3. ¬P → P
  4. P → ¬P
  5. Q ∨ ¬Q → P

H.   Some of the rules of inference introduced in Chapter 2 are tautological implications and some are tautological equivalences. Use truth tables to show which of the following are tautological implications and which are tautological equivalences:

1. Law of Simplification	(Show whether P & Q → P is a tautology and whether P & Q ↔ P is a tautology)
2. Law of Double Negation	(Show whether P → ¬¬P is a tautology and whether P ↔ ¬¬P is a tautology)

  3. Law of Addition
  4. Commutative Laws
  5. The new law: From P → Q we may infer ¬(P & ¬Q).

REVIEW TEST

I.   In the following examples, first translate into logical symbols using the given letters for the atomic sentences, then use truth diagrams to find the truth value of the molecular sentences from the given truth values of the sentences which are the parts.

B='Franklin was born before Washington' (true)

W='Washington was born in the eighteenth century' (true)

Q='John Quincy Adams was born before John Adams' (false)

L='Lincoln lived during the same period as Franklin' (false)

a. If Franklin was born before Washington, then John Quincy Adams was not born before John Adams.

b. If either Lincoln lived during the same period as Franklin or John Quincy Adams was born before John Adams, then Washington was born in the eighteenth century.

c. If Lincoln did not live during the same period as Franklin, then either Washington was not born in the eighteenth century or John Quincy Adams was born before John Adams.

II. Use truth tables to find the validity or invalidity of the following symbolized arguments:

a. $A \rightarrow B$

Therefore: $\neg B \rightarrow \neg A$

b. $x < 4$

Therefore: $x = y \ \lor \ x < 4$

c. $A \rightarrow (B \lor C)$

$C \ \& \ B \rightarrow C$

Therefore: $A \rightarrow C$

d. $A \lor B$

Therefore: A

III. Let A, B, and C be any three distinct atomic sentences. Decide by truth tables which of the following sentences are tautologies:

a. $A \lor \neg A$

b. $\neg A \ \& \ B \leftrightarrow (B \rightarrow A)$

c. $\neg(\neg A \lor B) \rightarrow \neg B \ \& \ A$

d. $A \rightarrow (\neg A \rightarrow B)$

e. $[(A \rightarrow B) \leftrightarrow B] \rightarrow A$

IV. Show by truth tables which of the following are tautological implications.

a. $\neg P \lor \neg Q \rightarrow \neg P \ \& \ \neg Q$

b. $[(P \rightarrow Q) \ \& \ (R \rightarrow P)] \rightarrow (R \rightarrow Q)$

V. Show by truth tables which of the following are tautological equivalences.

a. $(P \rightarrow Q) \leftrightarrow (\neg Q \rightarrow \neg P)$

b. $P \leftrightarrow P \lor Q$

c. $(P \rightarrow Q) \leftrightarrow \neg P \lor Q$

CHAPTER FIVE
TERMS, PREDICATES, AND UNIVERSAL QUANTIFIERS

▶ 5.1 *Introduction*

So far in our study of logical inference we have examined the logical form or structure of molecular sentences but we have not analyzed the logical structure of atomic sentences. We may ask ourselves this question: "Do the rules of inference we have been considering permit us to make all the inferences and derive all the conclusions we think are valid?" It is not difficult to find examples that show the answer to this question is "no". Consider the following argument.

> Premise: All birds are animals.
> Premise: All robins are birds.
> Conclusion: All robins are animals.

This certainly seems to be a correct argument. Let us write it in the following general form:

> Premise: All *B* are *A*.
> Premise: All *R* are *B*.
> Conclusion: All *R* are *A*.

Let *A*, *B* and *R* be any things that you may choose. Whenever the premises are true, you will find that the conclusion is true.

On the other hand, let us symbolize this argument as we have done in earlier chapters.

Let	P = 'All robins are animals'
	Q = 'All robins are birds'
	R = 'All robins are animals'.

Letting P, Q, and R stand for these sentences, the argument then becomes

P	Premise
Q	Premise
R	Conclusion

It is very clear that we cannot derive **R** from **P** and **Q** by the rules we have considered so far.

Apparently we need more rules of inference to do everything we want to do in logic. But before we introduce new rules we must look more carefully at the structure of atomic sentences.

E X E R C I S E 1

A.   In the following arguments both the premises and the conclusions are atomic sentences. Although the conclusions cannot be derived by the rules and methods you know now, some of the arguments are valid and others are not. Read the arguments and tell whether you think the conclusion follows or does not follow from the premises. (You are not required to do derivations here but simply to tell what *seems* logical to you.) A word of caution: Do not confuse factual truth and logical validity.

1. All frogs are amphibians.
   All amphibians are vertebrates.
   Therefore, all frogs are vertebrates.
2. Some students study logic.
   All students who study logic know the word 'premise'.
   Therefore, some students know the word 'premise'.
3. All of the trees in our yard lose their leaves in autumn.
   No pine trees lose their leaves in autumn.
   Therefore, some of the trees in our yard are pine trees.
4. All reptiles are cold-blooded animals.
   All snakes are cold-blooded animals.
   Therefore, all snakes are reptiles.
5. All Rick's friends are boys who are on the basketball team.
   All boys on the basketball team are tall boys.
   Therefore, all Rick's friends are tall boys.
6. Some of the figures on this paper are pentagons.
   All pentagons are five-sided figures.
   Some of the figures on this paper are five-sided figures.
7. Any object which emits light because of the energy of its particles is a luminous body.
   The moon is not a luminous body.
   Therefore, the moon is an object which emits light because of the energy of its particles.

8. Plato was a Greek philosopher.
   Some Greek philosophers were citizens of Athens.
   Therefore, Plato was a citizen of Athens.
9. Although rich in petroleum deposits, none of the Persian Gulf countries are highly industrialized countries.
   Iraq is a Persian Gulf country.
   Therefore, Iraq is not a highly industrialized country.
10. Some fractions are greater than some whole numbers.
    The number 8/2 is a fraction.
    The number 3 is a whole number.
    Therefore, 8/2 is greater than 3.
11. No dogs are amphibious.
    Snoopy is a dog.
    Therefore, Snoopy is not amphibious.
12. All members of the City Council live within the city limits.
    None of the Smiths live within the city limits.
    Therefore, none of the Smiths are members of the City Council.
13. None of today's bills is likely to pass.
    All of the bills likely to pass are supported by the President.
    Therefore, none of today's bills is supported by the President.
14. No crustacea belong to the phylum mollesca.
    All clams and snails belong to the phylum mollesca.
    Therefore, no clams or snails are crustacea.
15. All A is B.
    All B is C.
    Therefore, all A is C.
16. Some members of the Council are Democrats.
    Some Democrats are opposed to White's re-election.
    Therefore, some members of the Council are opposed to White's re-election.
17. Some A are B.
    All B are C.
    Therefore, some A are C.
18. All molecular sentences contain connectives.
    Some molecular sentences have just one atomic sentence.
    Therefore, all sentences which contain connectives have just one atomic sentence.
19. All molecular sentences contain connectives.
    Some molecular sentences have just one atomic sentence.

Therefore, some sentences which contain connectives have just one atomic sentence.

20. None of these ships are sailing ships.

Only sailing ships need the wind for propulsion.

Therefore, none of these ships needs the wind for propulsion.

B. Suggest a conclusion that you think would follow from the premises given in each exercise below, even though you are not yet able to prove the conclusion valid. (Again, you are asked to tell what *seems* to follow, not to derive a conclusion by formal methods.)

1. All the team members won their events.

All who won their events received medals.

2. No atomic sentences contain connectives.

This sentence is an atomic sentence.

3. Some mammals are herbivorous animals.

No herbivorous animals are meat eaters.

4. All of my friends will attend the meeting.

Mike is one of my friends.

5. None of the committee members is present.

All of the persons directly affected by the amendment are present.

▶ 5.2 *Terms*

We shall begin our analysis of atomic sentences by finding out what terms are. Consider the sentences,

Mary is absent.

John is slow.

This book is red.

Two is less than three.

In these sentences the terms are the words 'Mary', 'John', 'this book', 'two', and 'three'. These examples suggest a simple definition.

A *term* is an expression that either names or refers to a unique object.

Later we shall make this definition more complete, but this is the basic idea. Notice that a term need not be a simple name like 'Mary' or 'John'. A term may also be a phrase, like 'this book', '3 + 2' or 'the first President of the U.S.', which refers to a particular individual or thing. Some terms are names and some terms are descriptions referring to an individual or object. Although we shall not be too concerned with the distinction between names and descriptions in the sections to follow, it may be useful here to look at some further examples and distinguish the names from the descriptions.

Consider the sentences,

> Brazil is the world's largest producer of coffee.
> This book is too heavy.
> $1 + 1 = 2$.

In these sentences we have as names 'Brazil' and the Arabic numerals '1' and '2'. We have as descriptions the phrases 'the world's largest producer of coffee', 'this book', '$1 + 1$'. Note that we will consider the phrase '$1 + 1$' a *description* referring to the number 2 and the numeral '2' a *name* of the number 2. Of course, we have more than one name for the number two, for example in addition to the Arabic numeral '2', we have the English word 'two' and the Roman numeral 'II'. Like a name, a description which is a term identifies a particular person or thing.

EXERCISE 2

A.   List the terms in the following sentences.

1. This exercise is very easy.
2. China is the most populous country in the world.
3. The game will begin early.
4. $5 + 4 = 3 + 6$.
5. Seven is more than three plus three.
6. William Shakespeare is the author of *Macbeth*.
7. Two times three is less than seven times one.
8. Mike is the president of our class.
9. My mathematics grade has improved.
10. $2^3 = 8$.
11. Elizabeth II is the Queen of England.
12. Paris is the capital city of France.

B.   In the following sentences, list separately the terms which are names and those which are descriptions.

1. The African continent is larger than the North American continent.
2. The square root of 25 is 5.
3. John is the fastest runner on the squad.
4. C is more than XXXII.
5. 4 times 20 is less than 3 times 30.
6. Susan is today's discussion leader.
7. That ladder is very unsteady.
8. The treasurer of the senior class is Larry.
9. The Andes are the longest mountain chain in the world.
10. $11 + 11 = 10 + 12$.
11. That book is very informative.
12. The country to the north of the United States is Canada.

▶ 5.3 *Predicates*

Consider the sentence,

Socrates is wise.

From the discussion in Section 5.2 we know that 'Socrates' is a term. What about the phrase 'is wise'? It is not a term but it does tell us something about Socrates. The phrase 'is wise' is a predicate. Ordinarily in atomic sentences the subject of the sentence is a term and the predicate is the remainder of the sentence that says something about the subject. In the example, the term 'Socrates' is the subject of the sentence and the phrase 'is wise' is the predicate that says something about Socrates. Let us look at some more examples.

John is swimming.
Mary sings.
Susan is sad.
Bill runs quickly.

You can pick out the terms in these sentences. The predicates should also be clear. They are 'is swimming', 'sings', 'is sad', and 'runs quickly'.

Let us see how we would symbolize these sentences. Let 'S' be the predicate 'is swimming' and let $j = $ John. Then we may symbolize the sentence 'John is swimming' by

$$Sj.$$

Let '*F*' be the predicate 'sings' and *m* = Mary. Then we may symbolize the sentence 'Mary sings' by

$$Fm.$$

We use only a single letter for the whole predicate. Thus in the sentence 'Bill runs quickly' let '*R*' be the predicate 'runs quickly' and *b* = Bill. Then the sentence may be symbolized by

$$Rb.$$

EXERCISE 3

A.   What are the complete predicates in the following sentences?

1. Jane walks slowly.
2. The lecturer is speaking rapidly.
3. Tom scores.
4. The first game is ending quickly.
5. John is very intelligent.
6. Walking at the front is the chancellor.
7. Entering now is the judge.
8. Ann scores highest.
9. Bill is running faster.
10. The president is speaking.
11. Jack listens closely.
12. Susan rides.
13. The temperature climbs steadily.
14. Frank lives nearby.
15. Mary sings beautifully.

B.   Symbolize the following sentences:
Example: George is running.

> Let '*R*' be the predicate 'is running'.
> Let *g* = George.
> Then: *Rg*.

1. The light ray is refracted.
2. The crowd dispersed quietly.
3. A faint breeze stirred.
4. The Cheshire Cat grinned.
5. Mehitabel meows.

  6. Susan walks gracefully.

  7. Jack enters.

  8. Mr. Smith scowls.

  9. Mrs. Brown cooks.

10. Cathy is studying.

11. That painting is lovely.

12. John's horse jumps faultlessly.

13. Mr. White works nearby.

14. The sun was high overhead.

15. George is waiting patiently.

▶ 5.4 *Common Nouns as Predicates*

It sometimes takes a little care to distinguish terms from predicates. Consider

Socrates is a man.

We all know that 'Socrates' is a term. We might also tend to think that 'man' is a term but it does not identify a particular person or thing. In English grammar, the word 'man' is called a common noun just because it is not the name of some particular person or object. From the standpoint of logic it is convenient to let common nouns serve as parts of predicates; thus in the above sentence the phrase 'is a man' is the predicate, and the common noun 'man' by itself is not a term.

Some other sentences in which common nouns serve as parts of predicates are

Chicago is a city.
Einstein was a brilliant scientist.
Mars is a planet.

In these three sentences the common nouns 'city', 'scientist', and 'planet', are parts of the predicates.

In grammar it is useful to distinguish between predicates that are simply composed of a verb and those which have a more complicated structure because they use common nouns or other parts of speech. In logic this distinction is not important. What is important and what we must be clear about is the distinction between terms and predicates. Remember that terms are expressions that name or describe some unique object. Predicates, on the other hand, do not name objects but tell us something about them. It is natural to say that predicates also describe objects. We may be more precise about this point. In logic when we say that a term *describes* an object we mean that from the description we know exactly what the object is. In this sense a term *completely* describes the object. It identifies that particular object. Therefore, common nouns may also be part of terms when a particular thing is referred to. For example 'this girl' or 'the first boy in line'. Predicates, on the other hand, only partially describe objects. From the sentence 'Mary sings' we know something about Mary because of the predicate, but from the predicate alone we cannot tell the sentence is about Mary.

We now have two different parts of a sentence in our logical classification in which common nouns may occur. They may be used in constructing terms as in 'that man' or 'the building on the corner of 42nd St. and 5th Ave.'. And they may be used in constructing predicates as in 'is a diligent student' or 'is a tree'.

Several common nouns may be used to construct a term. For example, the term 'the man who robbed the bank' uses the common nouns 'man' and 'bank'.

EXERCISE 4

A.   Make up five atomic sentences in which the following common nouns occur *in the predicates:* game, girl, boy, school, building.

B.   Make up five more atomic sentences which use common nouns different from those in Exercise A and those in the examples given below.

C.   What are the common nouns in the following sentences?

1. Jay is a student.
2. Blue and yellow are complementary colors.
3. This sea anemone is an animal.
4. These redwoods are giant trees.
5. ¾ is a rational number.
6. Mary is a nurse.
7. Tom's father is an engineer.
8. Larry is a boy.
9. Mr. Jones is a salesman.
10. Mrs. Green is a history teacher.

D. List the sentences in which there are common nouns. Then tell which of these common nouns are parts of predicates. Finally, tell which common nouns are not parts of predicates but are used as terms, for instance, 'volumes' in (5).

  1. Jim acts very quickly.
  2. Tim is a man who likes books.
  3. Joe is reading rapidly.
  4. Jane appears to be concentrating.
  5. These volumes are reference books.
  6. His plan worked perfectly.
  7. Venus is a planet.
  8. That object is a distant star.
  9. That plant is a vegetable.
10. The clock ticks steadily.
11. This package contains clothing.
12. His package is being delivered now.
13. Your phone is ringing.
14. Sally is a high school student.
15. Carol is a pianist.

E. Which are the *terms* that are the subject and not part of the predicate in each sentence.

  1. Two plus two is equal to three plus one.
  2. Jack is a ballplayer.
  3. Susan is a secretary.
  4. Mary is waiting patiently.
  5. $5 \times 6 = 30$.
  6. This bird is a loon.
  7. This rose is a fragrant flower.
  8. Jane walks away.
  9. Three is a prime number.
10. John speaks distinctly.

F. Symbolize the following sentences. Use a capital letter to represent the entire predicate and a lower case letter for the term.

  1. Bill is a member of the club.
  2. Terry speaks softly.
  3. Charles is a musician.

4. This desk is cluttered.
5. This table is round.
6. The Hundred Year's War began in 1337.
7. Mr. James' car is a sports car.
8. Australia is a continent.
9. This block of brass has a mass, 500.0 grams.
10. Zero is a number.
11. Mercury is a liquid that expands in proportion to temperature increase.
12. This dinosaur lived during the Jurassic Period.
13. The Russian city of Sevastapol is a port on the Black Sea.
14. This circle has a tangent AB.
15. This book has a page missing.

G. In the sentences below two atomic sentences are joined by a connective. Symbolize the entire sentence using lower case letters for terms and capital letters for predicates.

1. If Jane is tall then Susan is small.
    Example: Let 'T' be the predicate 'is tall'
            'S' be the predicate 'is small'

$$j = \text{Jane}$$
$$s = \text{Susan}$$
    Then $Tj \rightarrow Ss$

2. Either Karen is late or Ruth is early.
3. Eubola is a Greek island and Crete is a Greek island.
4. If Tom is elected then George will be appointed.
5. If Judy is a senior then Bob is not a senior.
6. Either the train has been delayed or this schedule is wrong.
7. George Sands and George Eliot are not men.
8. Either the specific gravity of helium is less than 1.000 or the air will not push the balloon upward.
9. Buddhism is a religion that originated in India but it became more widespread in China.*
10. If Larry is here then this meeting can begin.
11. John will win if and only if he trains every day.
12. Sally will be a musician if and only if she practices diligently.

---

* The word 'but' is a connective which has about the same sense as 'and'. Its truth-functional analysis is exactly like that of 'and' and therefore in logic we will treat it as we do the connective 'and' and symbolize it by ' &'.

▶ 5.5 *Atomic Formulas and Variables*

In predicate logic the smallest expression that can make any sense standing alone is a predicate letter with a term attached. For example,

(1)   *Lj*

which might stand for the atomic sentence

(2)   James studies logic.

By itself 'studies logic' does not say anything, neither does 'James'; and neither does '*L*' or '*j*'. '*L*' is only a predicate and '*j*' is only a term.
    Now consider the expression

(3)   *x* is an even number,

which might be symbolized,

(4)   *Ex*.

The *forms* of (3) and (4) are the same as the forms of (2) and (1) respectively. But (3) and (4) do not talk about anything in particular and we cannot say they are either true or false because *x* is no object in particular. Nevertheless, we will find use for expressions like (4) as well as (1) standing alone (or as parts of larger expressions). We will call them atomic formulas.
    In (3) and (4) let us replace *x* by '4'. This gives us the true atomic sentences '4 is an even number' and '*E4*'. If we replace '*x*' by '5' we get false sentences. We could choose any of various terms that name or describe unique objects to put for *x* in (3) or (4): 6, 2+8, Philadelphia, etc. Each would result in true or false atomic sentences. When the letters '*x*', '*y*', '*z*' are used as terms, yet standing for no particular objects, we call them *variables*. In specific cases we may replace them by names of particular objects. So variables are considered terms too although not naming or referring to any unique object. This is the reason that when terms were introduced we said we would give a more complete definition later. That definition is

> A *term* is an expression which either names or refers to a
> unique object, or is a variable which may be replaced by
> an expression naming or referring to a unique object.

It is natural to ask what variables correspond to in grammar. It is clear they are not verbs, adjectives, adverbs, or nouns. But they correspond very closely to pronouns. Consider the following atomic formulas containing variables.

x is a man.	$x = y$
y is an astronaut.	$x > 1$
z is a book.	$y > 10$
u is a famous painting.	$x + y = z$
x is a new black sedan.	$x - 1 = 3$

If we write the five atomic formulas on the left replacing the variables by pronouns, we obtain

He is a man.
He is an astronaut.
It is a book.
It is a famous painting.
It is a new black sedan.

At first glance, we would be inclined to say that these sentences using pronouns are either true or false. However, the peculiar thing about pronouns, which makes them like variables, is that simply by looking at sentences in which they occur, we cannot tell if the sentences are true or false. With just the sentence 'He is a man' we do not know what unique object the pronoun 'He' refers to. To find the object to which the pronoun refers we must consider the full context or situation in which the sentence is used. Without this context we cannot say if the sentence 'He is a man' is true or false. These same remarks apply to the other examples.

EXERCISE 5

A.   List the pronouns in the following sentences.

1. She is Terry's friend.
2. It is an insect.
3. He works at the supermarket.
4. They are senators.
5. It is our favorite song.
6. She is a secretary.
7. She writes books.

8. He is a doctor.

9. It is a large house.

10. It is an early-blooming plant.

B.   In the examples of Exercise A replace the pronouns by variables. Use '*x*', '*y*', and '*z*' for variables.

We can now give a concise definition of an atomic formula.

An *atomic formula* is a single predicate with the appropriate number of terms attached.

Until now we have considered only predicates for which a single term is appropriate. They are called one-place predicates. Consider the following:

(1)   Mr. Smith is the father of Jimmy.

If we let '*s*' be 'Mr. Smith' and '*K*' be the one-place predicate 'is the father of Jimmy' then (1) would be symbolized

$$Ks.$$

But suppose (1) were a premise and there were a second premise saying

(2)   Anyone whom Mr. Smith is the father of has black hair.

From (1) and (2) we could logically conclude that Jimmy has black hair. But this could not be done with '*K*' as the one-place predicate 'is the father of Jimmy'. So let us use the predicate '*F*' to stand for 'is the father of' and '*j*' for 'Jimmy'. We can then symbolize (1) as the atomic formula

$$Fsj$$

'*F*' is a two-place predicate because the appropriate number of terms to attach to it is two. '*Fs* ' or '*F j*' would stand for 'Mr. Smith is the father of' and 'is the father of Jimmy'. Neither of these is complete. They cannot stand alone, so they are not atomic formulas.

E X E R C I S E 6

A.   '*B*' is a one-place predicate, '*D*' is a two-place predicate, and '*G*' is a three-place predicate. For each of the following tell whether or not it is an atomic formula.

1. *Ba*

2. *¬Dde*

3. *Bx* →

4. *Dxy*

5. *Gabc*

6. *Dae* ∨ P

7. *Dcy*

8. *Gxyz*

9. ↔

10. *Bc* & *Dab*

11. *Gaxy*

12. *Bz*

B.   Make up five atomic formulas using variables '*x*', '*y*', and '*z*'.

C.   Which of the following are atomic formulas?

1. $x = y$.

2. $z < x$.

3. $x = y$ and $y - 1 = z$.

4. $x + 1 = y + 1$.

5. $x + y = y + w$.

6. If $z < x$ then $z < y$.

7. If $z < x$ then $y > z$.

8. $x + y > z$.

9. If $y - 1 = z$ then $x - 1 = z$.

10. Either $x > y$ or $z > x$.

11. $x \neq z$.

12. $x + z = y + z$.

13. $x - w \neq y + z$.

14. $x > w$ and $y < z$.

15. $x = y \rightarrow z \neq y + 1$.

16. $x = 2 \times 2$.

17. $y = 2 + 2$.

18. $x = 3 + 1$ and $y \neq 3 + 1$.

19. $x > 1$.

20. $z > 2$.

An acquaintance with variables and atomic formulas makes it possible to give a clear form for translation from English to the symbolism of predicate logic. Consider the example,

Eisenhower appointed Chief Justice Warren.

Axy   ↔   *x* appointed *y*

e = Eisenhower

w = Chief Justice Warren

In symbols: *Aew*

When we use equivalences like '*Axy*   ↔   *x* appointed *y*' we are saying that whenever we have 'appointed' in English we will use '*A*' in symbols and whenever we have '*A*' in symbols it means 'appointed' in English. '*x*' and '*y*' are used to show that 'appointed' and '*A*' are two-place predicates requiring two terms. The terms '*e*' and '*w*' are not used in giving the translation of 'appointed' because the translation of each term and each predicate must be given separately to keep them clear. We separately give translations for 'Eisenhower', 'Chief Justice

Warren', and 'appointed', but not for 'Eisenhower appointed Chief Justice Warren' all at once.

Notice the following. (1) The symbols for terms and for predicates are given separately. (2) The number of terms appropriate to the predicate is indicated by attaching that many variables to the predicate. In other words, the translation of a predicate is shown by showing the symbol for the predicate of ordinary language as part of an atomic formula, in this case '*Axy*'. (3) The order in which we write down terms attached to predicates of more than one place is important. (4) An equivalence connective, ↔, is put between the atomic formula in logical symbols and its English equivalent. (5) An equal sign, =, is put between the logical symbols for terms standing for unique objects and the English for the same object. (6) The logical symbols are given on the left, the English words on the right.

We will always use this form when giving translations. Telling what logical symbols will be used for English atomic sentences, predicates, or terms can be called *defining symbols*.

It is also acceptable to write '*xAy*' instead of '*Axy*'. Thus, the atomic sentence above, '*Aew*', could also be symbolized '*eAw*'. In either case the order of the two terms must be the same.

EXERCISE 7

Translate the following into the full form of predicate logic, first defining symbols then giving the translation.

1. The Washington Monument is 555 feet tall.
2. Aristotle was the teacher of Alexander the Great.
3. The senate has one hundred members.
4. Lindbergh made the first solo flight to Paris.
5. Fujiyama is a very beautiful mountain.
6. The highest falls in Yosemite Park drop 2,526 feet.
7. The United States gets much coffee from Brazil.
8. The Golden Gate Bridge is red.
9. The real number system is a field.
10. Triangle *ABC* is congruent to triangle *DEF*.

Atomic formulas whose terms do not use variables are atomic sentences. With connectives they form molecular sentences just as in sentential logic. Atomic formulas containing variables also can be combined with connectives, but the results are not true or false sentences. In predicate logic, expressions containing connectives are called *molecular formulas* whether they contain variables or not.

Consider the following examples:

Example a.

> If Michelangelo was a Renaissance artist then Leonardo daVinci was a Renaissance artist.

$$Rx \quad \leftrightarrow \quad x \text{ was a Renaissance artist}$$
$$m = \text{Michelangelo}$$
$$l = \text{Leonardo daVinci}$$

In symbols: $Rm \quad \rightarrow \quad Rl$

Example b.

> Bill helps Harry and is helped by Jack.

$$Hxy \quad \leftrightarrow \quad x \text{ helps } y$$
$$b = \text{Bill}$$
$$h = \text{Harry}$$
$$j = \text{Jack}$$

In symbols: $Hbh \quad \& \quad Hjb$

Notice that the right member of the conjunction, 'Bill is helped by Jack' is treated like 'Jack helps Bill'.

Example c.

> If x is greater than two and two is greater than z then x is greater than z.

This can be translated using standard mathematical and logical symbols. When this is done it is unnecessary to define the symbols. So we immediately write the molecular formula,

$$x > 2 \quad \& \quad 2 > z \quad \rightarrow \quad x > z$$

EXERCISE 8

A. Tell which of the following could only be translated as atomic formulas and which could be translated as molecular formulas.

1. Dave's income increased and the prices of consumer goods increased.
2. If Dave's income did not go up proportionately then his real income decreased.
3. y is the consumer price index for the month.
4. Either real income increases or the standard of living does not go up.
5. x travels northward.
6. x is 1,000 miles from z.
7. z travels at the rate of 300 miles per day.
8. $\frac{5}{6}$ is not in the set of whole numbers.
9. x is a whole number.
10. z is not a rational number.
11. Jean is not a sophomore.
12. The moon is the only natural satellite of the earth.
13. z is the number of juniors in the school.
14. Don pulled with a force of 80 pounds and Paul pulled the other way with a force of 90 pounds.
15. y is the magnitude of the resultant forces.

B. Give a predicate translation of each of the following, first defining your symbols.

1. If the Nautilus is in equilibrium then either it is at rest or it is moving at constant speed in a straight line.
2. Martha loves Don and Don loves Martha.
3. Either Mrs. Clark gives Walter a ride or he is late to his appointment.
4. If the elastic limit of the spring is exceeded then its molecular forces are overcome and the spring does not return to its original form.
5. If either Joe is not the brother of Maria or Maria is not the sibling of Joe then Mr. Lopez is not the father of Joe and Joe is the cousin of Maria.

▶ 5.6 *Universal Quantifiers*

Let us look at the atomic formula,

$$x \text{ is tall.}$$

If we replace 'x' by 'Lincoln' we obtain the true atomic sentence

'Lincoln is tall'. If we examine the atomic formulas given as examples in the last section, we can see in every case that if we replace the variables 'x', 'y', and 'z' by terms referring to some unique object we get an atomic sentence that is either true or false.

Is there any other way to change atomic formulas into true or false sentences? The answer is "yes". Instead of putting 'Lincoln' for 'x' we may say, 'Everything is tall'. This is a grammatical sentence which is false. Instead of writing 'Everything is tall' we can write the sentence as 'Every x is tall'. In logic we usually express the sentence in this way:

For every x, x is tall.

The phrase 'For every x' is a *universal quantifier*. It is called a universal quantifier because it uses the variable 'x' to assert that everything in the universe has a certain property; in the present case, the property of being tall.

As another example consider the atomic formula of arithmetic

$$x > 0.$$

We can make this atomic formula true or false by replacing 'x' by the name or description of some particular number. For example, if we replace 'x' by '$1+1$', we obtain the atomic sentence

$$1+1 > 0,$$

which is true. If we replace 'x' by '-3', we obtain the false atomic sentence,

$$-3 > 0.$$

If we add a universal quantifier to the atomic formula '$x > 0$', we obtain the false sentence,

For every x, $x > 0$.

This sentence is false because, for example, if $x = -1$ then x is not greater than 0. In other words, the sentence asserts that *every* number is greater than 0, but the number -1 is not, and so the sentence is false.

The symbol for the universal quantifier is an upside-down '**A**'; we symbolize the last displayed sentence by

$$(\forall x)(x > 0).$$

The choice of the letter 'A' turned upside-down comes from another common idiom of universal quantification. Instead of saying 'For every x, $x > 5$', we could say

For all x, $x > 5$.

In either case it would be symbolized

$$(\forall x)(x > 5).$$

From a logical standpoint the phrase 'For all x' is used in the same sense as the phrase 'For every x'. As we shall soon see, the shift from 'every' to 'all' usually involves a shift from the singular to the plural in English, but this is a superficial change similar to the one noted earlier for common nouns. It does not reflect a logical shift. 'All cats have claws' is translated in the same way as 'Every cat has claws'.

The phrase 'everyone' is also a common idiom for expressing universal quantification. In ordinary language, rather than saying 'For every x, x is wise' we would say 'Everyone is wise'. The translation would be

$$Wx \quad \leftrightarrow \quad x \text{ is wise.}$$

In symbols:

$$(\forall x)(Wx).$$

The following list summarizes the more common idioms used to express a universal quantifier:

For every x . . . . . .
For everything . . . . .
Everyone . . . .
For all x . . . .
All things . . . .
Any . . . .

Notice that in the sentences 'For every x, $x > 0$' and 'For all x, $x > 5$' we take it for granted that x is a number. The universal quantifier in these cases does not refer to all things or entities but only to all numbers. Often we are interested not in *all* things in the universe but in some definite set of things. In the example just given we considered formulas of arithmetic and therefore we were only concerned with the set of numbers. The set of things being considered in a discussion is called the *domain of individuals* or the *domain of discourse*. Thus in some examples we restrict the domain to a particular set and then the universal quantifier refers to every element in that set. In other examples we do not restrict the domain but let the universal quantifier cover all entities or things in the universe.

Usually the context of the discussion makes clear the domain. For instance, the use of mathematical symbols in many of our examples will indicate that the domain is the set of numbers. Thus, as we have seen, the universal quantifier 'For every x' means that we are asserting something for every element in that domain (in other words, for every number). Sometimes a particular English idiom used to express universal quantification indicates the domain. The words 'everyone' or 'everybody', for example, suggest that the domain of individuals is the set of human beings. In these cases, the universal quantifier 'for every x' refers to every human being.

We can confine ourselves to a particular limited domain in any sentence. In a sentence such as 'For every x, $x>0$' it must be understood that the domain is restricted to the set of numbers. If the domain were not restricted, we would have to make the sentence the conditional, 'For every x, if x is a number then $x>0$'. In particular discourse, as in the examples from arithmetic, it is much more convenient to confine ourselves to a restricted domain because the subject of that discourse is limited to a fixed domain like the set of numbers.

EXERCISE 9

A. Convert each of the atomic formulas into a true or false atomic sentence by replacing the variables by names or descriptions of unique objects. State whether the resulting sentence is true or false.

1. x is a United States Senator.
2. z is a teacher.
3. y is a good book.
4. z is a number greater than 4.
5. x is a friendly person.
6. z is the best logician in the room.
7. z plays baseball.
8. y is a flower.
9. x is the principal of the school.
10. x is an astronaut.
11. z is the Secretary of State.
12. y is the President of the United States.
13. x is the date of my birthday.

14. *y* signs all paper currency.

15. *z* is a member of the President's Cabinet.

B. Convert the atomic formulas of Exercise A into true or false sentences by adding universal quantifiers using the logical symbol ∀. State for each of the resulting sentences whether it is true or false.

Certain idioms of universal quantification are used to express negation at the same time. Consider the example,

No one likes poisonous mushrooms.

The word 'No' does double duty both as a universal quantifier and as an expression of negation. Using the variable '*x*', we may translate this sentence.

For all *x*, *x* does not like poisonous mushrooms

or

($\forall x$)(*x* does not like poisonous mushrooms).

The sense in which 'No' in the original sentence expresses universal quantification is now expressed by the phrase 'All *x*' and the sense in which the word 'No' in the original sentence expresses negation is now expressed by the word 'not'. In order to symbolize completely our translated sentence, we first define,

Lx ↔ *x* likes poisonous mushrooms

and then symbolize the sentence by

($\forall x$)($\neg Lx$).

Still another idiom that expresses *both* universal quantification and negation is shown in the following sentence:

Nothing is absolutely bad.

In this sentence the word 'Nothing' serves both as a universal quantifier and as an expression of negation. Again it may help to add variables in the English first:

For all *x*, *x* is not absolutely bad.

Then, define:

Bx ↔ *x* is absolutely bad.

The sentence is symbolized

$$(\forall x)(\neg Bx)$$

The following list summarizes the more common idioms used to express both a universal quantifier and a negation:

No one .　.　.　.　.　.
None　.　.　.　.　.
Nothing　.　.　.　.　.
No　.　.　.　.　.　.　.

EXERCISE 10

A. Use the universal quantifier to symbolize the following sentences, but do not use letters to stand for predicates; express the negations and predicates in ordinary English with variables.

 1. For every x, x has a name.
 2. Everything is subject to change.
 3. For all x, x is the value of a variable.
 4. Nothing is absolutely cold.
 5. For every y, y belongs to a set.
 6. Nothing changes.
 7. For all y, y is not perfect.
 8. All things are atoms.
 9. For all y, y is matter.
 10. For all y, y is an Idea.

B. In each of the following sentences the domain of individuals is the set of numbers. Use the universal quantifier to symbolize the sentences but do not use letters to stand for predicates. Express the negations and predicates with variables in standard mathematical symbols or ordinary English.

 1. For all z, $z > 0$.
 2. For every x, $x < x+1$.
 3. For every x, x is not divisible by zero.
 4. For all w, $w+0 = w$.
 5. For all x, x is not greater than x.

C. In each of the following sentences the domain of individuals is the set of human beings. Use the universal quantifier to symbolize the sentences but do not use letters to stand for predicates. Express the negations and predicates in ordinary English with variables.

1. No one welcomes disaster.
2. No one hears those sounds.
3. No one likes to be wrong.
4. No one is perfect.
5. Everyone has certain minimum food requirements.

It is necessary to distinguish the cases in which a negation follows the quantifier from the cases in which the negation precedes the quantifier. Consider the following sentence.

$$(1) \quad \text{Not everything is beautiful.}$$

This is simply the negation of

$$(2) \quad \text{Everything is beautiful.}$$

On defining: $Bx \leftrightarrow x$ is beautiful, sentence (2) is symbolized

$$(\forall x)(Bx)$$

and (1) is the negation of this

$$\neg(\forall x)(Bx).$$

English is sometimes not as precise as logical symbolism. There is a rich variety of ways to state a given idea in English. A number of different English forms may have the same meaning. On the other hand, one English sentence may sometimes have several different meanings. For example the place where 'not' appears in English sentences does not always tell us what to negate in the symbols. Consider

$$(3) \quad \text{Everyone is not a fool.}$$

At first it might seem that 'being a fool' is what is negated. Adding variables, the English would then be

For all x, x is not a fool,

which means

$$(4) \quad \text{No one is a fool.}$$

On defining: $Fx \leftrightarrow x$ is a fool, in symbols the sentence would be

$$(5) \quad (\forall x)(\neg Fx).$$

But if it means

$$(6) \;\; \text{Not everyone is a fool}$$

adding variables would give

$$\text{Not for all } x, \; x \text{ is a fool}$$

or in symbols

$$(7) \;\; \neg(\forall x)(Fx).$$

Sentence (3) is ambiguous and it is necessary to consider what meaning the sentence has in each context. On the other hand, the choice of language in sentence (6) or sentence (4) is better because the logical meaning is clear. In sentence (6) the whole atomic sentence 'everyone is a fool' is negated. In sentence (4) the atomic formula 'Fx' is negated.

EXERCISE 11

A. Translate completely into the symbolism of predicate logic all the examples in Exercise 10 (A, B, and C). In Exercise 10 the universal quantifiers were added. Now complete the symbolization by using letters to stand for the predicates and adding the appropriate negation symbols.

B. Translate completely into the symbolism of predicate logic. It may help to put in variables in the English first if they are not already there, although it is not necessary to write this step. In some of the examples it is obvious that the domain of individuals is restricted. In each example in which there is a restricted domain indicate that domain.

Example:	Everything is good.
On putting in variables:	For all x, x is good.
On defining symbols:	$Gx \;\leftrightarrow\; x$ is good.
In symbols:	$(\forall x)(Gx)$

1. For every z, z is living.
2. Everyone wishes for good luck.
3. For all x, $x > x - 1$.
4. Everybody isn't right handed.
5. For every y, $y = y$.
6. For every z, z is a number.
7. Nothing is impossible.
8. No one enjoys defeat.

9. For all x, x is not absolutely stable.
10. Not everything is worth trying.
11. No one is omniscient.
12. All things have value.
13. For every x, x is wise.
14. For all y, y is foolish.
15. Everybody doesn't have two good eyes.
16. Everything is relative.
17. For every w, w is a man.
18. Everything has a history.

▶ 5.7 *Two Standard Forms*

We now turn to the symbolization of certain standard kinds of sentences that occur repeatedly in deductive arguments or in other scientific contexts. Each of these sentences uses a universal quantifier.

To begin with, let us consider the sentence,

(1) Every man is an animal.

According to our previous discussion of common nouns and predicates, from the logical standpoint there are two common nouns. Neither of them is used in constructing a term and therefore there are two predicates in this sentence, namely the predicate 'is a man' and the predicate 'is an animal'. We use these two predicates to translate the sentence as

(2) For every x, if x is a man, then x is an animal.

The important and interesting thing about the translation of (1) as (2) is that in sentential logic (1) would be translated as an atomic sentence, whereas (2) uses the sentential connective 'if . . . then . . .'. The basis for the change is this. If the common nouns are to be treated as predicates, then sentences like (1) cannot be translated as atomic formulas for an atomic formula can have exactly one predicate. By using common nouns rather than complete predicates, (1) expresses a relation between men and animals in the form of an atomic sentence. When this relationship is to be expressed by complete predicates, so that we can symbolize it, it is necessary to use a sentential connective to express it. That is what is done in sentence (2).

It is of the utmost importance to realize that the only sentential connective that will do in (2) is 'if . . . then . . . '. Suppose, for instance, that instead of using 'if . . . then . . .' as the sentential connective, we used 'and'. We would then translate (1) by the sentence,

(3) For every x, x is a man and x is an animal.

It should be apparent that the meaning of (3) is different from the meaning of (1). Just how does the meaning of the two examples differ? Sentence (3) says that everything is both a man *and* an animal, and this sentence is clearly false. It is not difficult to find examples which make (3) false. For instance, the page on which you are reading these sentences is certainly neither a man nor an animal, contrary to what sentence (3) asserts. Sentence (1), on the other hand, simply says that for everything we can think of, *if* it is a man then it must also be an animal. In our usual language this means the same as the true sentence 'Every man is an animal'.

On defining, $Mx \leftrightarrow x$ is a man, $Ax \leftrightarrow x$ is an animal, and also using the $\rightarrow$ sign and the symbol for the universal quantifier, we may symbolize (1) and (2),

(4) $(\forall x)(Mx \rightarrow Ax)$.

Sentence (4) exemplifies the standard form for sentences of the type 'Every such-and-such is so-and-so'.

Because logically 'every' and 'all' have the same force, it exemplifies just as well sentences of the type 'All such-and-such's are so-and-so's'. From a logical standpoint we could have replaced (1) by the sentence

All men are animals

without changing the logical force of what we were saying. Notice that the shift in grammar from the singular to the plural verb is of no logical importance in this case.

We now consider a case in which we have both universal quantification and negation. To exemplify the standard form we may begin with the sentence

(5) No man is immortal.

We translate this sentence as

(6) For all x, if x is a man, then x is not immortal

and (6) is translated into symbols by

$$(7) \quad (\forall x)(Mx \ \rightarrow \ \neg Ix),$$

where $Ix \ \leftrightarrow \ x$ is immortal.

Sentence (7) exemplifies the standard form for sentences of the type 'No such-and-such is a so-and-so' or in plural terms 'No such-and-such's are so-and-so's'.

In some cases a whole sentence like (4) is negated. Consider

(8) Not all women have long hair.

This is simply the negation of

All women have long hair.

On defining

$$Wx \ \leftrightarrow \ x \text{ is a woman}$$
$$Lx \ \leftrightarrow \ x \text{ has long hair}$$

Sentence (8) is symbolized

$$\neg(\forall x)(Wx \ \rightarrow \ Lx).$$

There are many different ways of saying the same thing in English. It is impossible to give standard forms for all of them. It is often necessary to consider just what the English says, or to try saying the same thing differently in English until you get it in a form in which you can clearly recognize the logical structure. If you get it in the form of a conditional, 'if . . . then . . .' the structure is usually the clearest. Consider the example

$$(\forall x)(Tx \ \rightarrow \ Px)$$

On defining

$$Tx \ \leftrightarrow \ x \text{ is a tree}$$
$$Px \ \leftrightarrow \ x \text{ is a plant}$$

any of the following are ways it might appear in English.

> Anything that is a tree is a plant.
> The only things that are trees are also plants.
> Only plants are trees.
> Nothing is a tree unless it is a plant.
> If something is a tree then it is a plant.

To summarize, sentences of the form 'All A's are B's' we symbolize as

$$(\forall x)(Ax \;\rightarrow\; Bx).$$

Sentences of the form 'No A's are B's', we symbolize by

$$(\forall x)(Ax \;\rightarrow\; \neg Bx).$$

EXERCISE 12

A.   Symbolize completely:

1. All sparrows are birds.
2. All pines are evergreens.
3. Every clam is a bivalve.
4. All Frenchmen are Europeans.
5. Every millionaire has wealth.
6. All fruit is delicious.
7. All grass is green.
8. All ice is cold.
9. Every stream runs downhill.
10. All horses are quadrupeds.

B.   Symbolize completely:

1. No peaches are vegetables.
2. No lemons are sweet.
3. No man is an island.
4. No cats are canines.
5. No adults are minors.
6. No foxes are foolish.
7. No fifth-grader is an infant.
8. No tyrant is a just man.
9. No birds are quadrupeds.
10. No fairy tales are true histories.

C.   Symbolize completely:

1. No canines are birds.
2. No abalones are bivalves.
3. No automobiles are jet-propelled.
4. No clowns are happy men.
5. All gloves have fingers.

6. All students are scholars.
7. No cooks are thin people.
8. All caves are shelters.
9. No turtles are sprinters.

D.   Symbolize completely the following sentences. Note that 'only' is another common idiom for expressing a universal quantifier.

1. Only protoplasm is living substance.
2. Only men are rational.
3. Only Europeans are Frenchmen.
4. All birds and fish are animals.
5. All horses and cows are quadrupeds.
6. All maples and oaks are deciduous trees.
7. Not all men are intelligent.
8. Not all men are honest.
9. Not all grass is green.

E.   Symbolize the quantifiers and sentential connectives, but retain the mathematical symbols.

1. For all x, if $x > 2$, then $x > 1$.
2. For all x, $x + 0 = x$.
3. For all x, if $x \neq 0$, then $x/x = 1$.
4. For all y, $y - y = 0$.
5. For all y, $y - 0 = y$.

R E V I E W   T E S T

I.   Make a list of the terms in the following sentences, then list the predicates in each.

a. Jim Taylor is a fullback.
b. $5^2 = 25$.
c. The big black bear lumbered slowly toward us.
d. Lincoln was the sixteenth President of the United States.
e. Two is the cube root of eight.

II.   Symbolize completely:

a. One is the reciprocal of one.
b. The Sixteenth Amendment permits federal income tax.
c. The Attorney General is appointed.

d. That book is a biography.

e. This book is a collection of essays.

f. The natural number system has a zero element.

g. Mrs. Costello is the mother of Dan.

h. Senators were not elected by direct popular vote before 1913.

i. If Ed is not a council member then Nick is not a council member.

j. Either $2+2\neq5$ or $2+3\neq6$.

k. Ruth is president and Theresa is treasurer.

l. The Bears will win if and only if George can play.

III. Fill in each blank with a single word.

a. Variables correspond to _____ in grammar.

b. An atomic formula may contain _____.

c. Atomic sentences do not contain _____.

IV. Make up five atomic formulas using variables.

V. Make the following sentences *true* sentences by replacing the variables by terms.

a. *x* is not in this class.

b. *y* is a number greater than ten.

c. *z* is the governor of this state.

d. *x* is not a positive number.

e. *y* is not the principal of this school.

VI. Completely symbolize the following sentences:

a. For all *y*, *y* equals *y* and *y* is not greater than *y*.

b. Everything has been said.

c. No man is both foolish and fair.

d. No number is both even and odd.

e. All men are mortal.

f. Everyone loves the circus.

g. No man is either totally wise or totally stupid.

h. Everything is either changless or changing.

i. For all *x*, *x* is positive if and only if *x* is greater than zero.

j. No music is noise.

k. Only positive numbers are greater than zero.

l. Not all numbers are positive.

m. Nothing is both round and square.

CHAPTER SIX
UNIVERSAL SPECIFICATION AND
LAWS OF IDENTITY

▶ 6.1 *One Quantifier*

Within the framework of sentential inference we cannot show that the following is a valid inference.

> Every citizen of California is a citizen of the United States.
> Governor Brown is a citizen of California.
> Therefore, Governor Brown is a citizen of the United States.

We have added to our set of logical tools the notation for universal quantifiers. Let us see what we can do by symbolizing the sentences of this argument in our new notation of quantification and then adding to our rules of inference a rule for dropping universal quantifiers. Our new methods permit us to analyze and use the detailed structure of the sentences in the argument.

On defining: $Cx \leftrightarrow x$ is a citizen of California, and $Ux \leftrightarrow x$ is a citizen of the United States, and $b =$ Governor Brown, we may symbolize the premises and conclusion of this argument by

> Prove: Ub
> (1) $(\forall x)(Cx \rightarrow Ux)$
> (2) Cb

This symbolization is the first step in our strategy of proof. The second step is to specify some particular object for x. The *intuitive* idea of specifying is that whatever is true for *every* object is true for any object we might wish to select—Governor Brown, for instance.

Once the quantifiers are gone, we are in a position simply to apply the sentential methods of derivation developed in Chapter 2.

The three-step strategy we follow is

Step 1. Symbolize premises.
Step 2. Specify objects to eliminate quantifiers.
Step 3. Apply sentential methods of inference* to derive
a conclusion.

Notice in the following proof that performing Step 2 changes the atomic formulas, 'Cx' and 'Ux', into atomic sentences. This is the reason that Step 3 can be taken.

Let us apply these three steps to our example. We obtain the derivation,

(1) $(\forall x)(Cx \rightarrow Ux)$ P
(2) Cb P
(3) $Cb \rightarrow Ub$ Specify b for x
(4) Ub PP 2, 3.

The rule that permits us to specify is called the rule of *universal specification*. We may drop the quantifier and replace its variable by *any* one term, thereby *specifying* any one individual we choose.* The idea is that any sentence which is true for *everything* must be true for whatever specified individuals we may choose. From the universal statement that something is true for anything we may choose we infer that the statement is true for some specific thing or things that we choose.

As a second example we may consider a simple argument about numbers.

Every positive number is greater than 0.
1 is a positive number.
3 is a positive number.
Therefore, 1 and 3 are greater than 0.

Define: $Px \leftrightarrow x$ is a positive number. Then, using the ordinary symbol $>$ for 'greater than' and the Arabic numerals as names of the numbers one and three, we may symbolize the argument,

Prove: $1>0$ & $3>0$
(1) $(\forall x)(Px \rightarrow x>0)$
(2) $P\ 1$
(3) $P\ 3$

---

* Later we shall state a necessary restriction to universal specification.

The application of universal specification to write a derivation of this argument is simple and easy.

$$(1) \ (\forall x)(Px \ \rightarrow \ x > 0) \qquad P$$
$$(2) \ P \ 1 \qquad P$$
$$(3) \ P \ 3 \qquad P$$
$$(4) \ P \ 1 \ \rightarrow \ 1 > 0 \qquad 1/x \quad 1$$
$$(5) \ 1 > 0 \qquad PP \ 2, 4$$
$$(6) \ P \ 3 \ \rightarrow \ 3 > 0 \qquad 3/x \quad 1$$
$$(7) \ 3 > 0 \qquad PP \ 3, 6$$
$$(8) \ 1 > 0 \ \ \& \ \ 3 > 0 \qquad A \ 5, 7$$

Notice that in lines (4) and (6) we have explicitly indicated the specification by '$1/x$' and '$3/x$' followed by the number of the line to which the universal specification was applied. In the future we shall use this slant line to show the specification. They are read, 'Putting 1 for x' and 'Putting 3 for x', and this leaves it understood that Universal Specification (US) is the rule that is applied. Notice that two different universal specifications may be applied to line (1). There is no limit to the number of times specifications may be applied to the same universal sentences, just as there is no limit to the number of times any line of a derivation may be used in deriving following lines.

EXERCISE 1

A. Symbolize the following premises and conclusions. Each example includes a term. Use lower case letters to symbolize terms.

1. All dogs are animals.
   Lassie is a dog.
   Therefore, Lassie is an animal.
2. No President of the United States was an immigrant.
   John Quincy Adams was a President of the United States.
   Therefore, John Quincy Adams was not an immigrant.
3. Every even number is divisible by two.
   Ten is an even number.
   Eight is an even number.
   Therefore, eight and ten are divisible by two.
4. No number is greater than itself.
   Three is a number.
   Therefore, three is not greater than three.

5. For every x, if x is a number then x plus one is greater than x.
   Four is a number.
   Therefore, four plus one is greater than four.
6. All parrots are birds.
   All birds are vertebrates.
   Polly is a parrot.
   Therefore, Polly is a vertebrate.
7. No fractions are integers.
   Four is an integer.
   Therefore, four is not a fraction.
8. All negative numbers are less than zero.
   Six is not less than zero.
   Therefore, six is not a negative number.
9. All presidents are elected Heads of State.
   No elected Head of State is a monarch.
   King Baudouin is a monarch.
   Therefore, King Baudouin is not a president.
10. No odd number is divisible by two.
    Six is divisible by two.
    Eight is divisible by two.
    Therefore, it is not the case that either six or eight is an odd number.
11. All congressmen are either senators or members of the House of Representatives.
    Mr. Anderson works in Washington but he is not a member of the House of Representatives.
    Therefore, if Mr. Anderson is a congressman he is a senator.

B.   Show a full derivation for the examples of valid inference given in Exercise A.

C.   In each of the following, derive the conclusion from the given premises using the complete standard form for proofs.

1. Derive: Pb
   (1) $(\forall x)(Px \quad \& \quad Rx)$

2. Derive: $Ft \quad \& \quad Fr$
   (1) $(\forall y)(Gy \quad \& \quad Fy)$

3. Derive: *Ga*
   (1) $(\forall x)(Hx \;\rightarrow\; Gx)$
   (2) *Ha*

4. Derive: *Fd*
   (1) $(\forall y)(Fy \;\leftrightarrow\; Hy)$
   (2) *Hd*

5. Derive: $2>0$
   (1) $(\forall x)(x>1 \;\rightarrow\; x>0)$
   (2) $2>1$

6. Derive: $3>0 \quad \& \quad 4>0$
   (1) $3>1$
   (2) $(\forall y)(y>1 \;\rightarrow\; y>0)$
   (3) $4>1$

7. Derive: $\neg Hf$
   (1) $(\forall x)(Hx \;\rightarrow\; Rx)$
   (2) $\neg Rf$

8. Derive: *Gb*
   (1) $(\forall x)(Gx \;\vee\; Jx)$
   (2) $\neg Jb$

9. Derive: *Jb*
   (1) $\neg Hb$
   (2) $(\forall y)(\neg Jy \;\rightarrow\; Hy)$

10. Derive: *Hx*
    (1) $(\forall x)(Gx \;\rightarrow\; Jx \;\&\; Hx)$
    (2) *Gx*

We may also apply universal specification to make inferences in which numbers or other objects are referred to by complex terms like '1 + 1' or '5 + 1' rather than by the simple name of the number. To construct an example we may use the first premise used in our last example.

> Every positive number is greater than 0.
> 1 + 3 is a positive number.
> Therefore, 1 + 3 is greater than 0.

Using the symbols already introduced, we may write the proof of this argument, applying universal specification to obtain line (3).

(1) $(\forall x)(Px \;\rightarrow\; x>0)$ P
(2) $P\;(1+3)$ P
(3) $P\;(1+3)\;\rightarrow\;1+3>0$ $1+3/x$ 1
(4) $1+3>0$ PP 2, 3

(We add parentheses in '$P\;(1+3)$' to make it immediately clear that the predicate 'P' applies to '$1+3$' and not simply '1'.)

The *operation signs* $+$, $-$, $\times$, and $\div$ make new terms from other terms. These particular signs indicate *binary* operations because each of them combines *two* terms to make *one* new complex term. For example, '3' is a term, '4' is a term, and '3×4' is a term. It is a complex term referring to the same number referred to by the term '12'. An operation sign that makes another term from only one term indicates a *unary* operation. Squaring a number is an example; '3' is a term and '3^2' is a term.

Of course, numerical operation signs are attached to terms, never to sentences. It makes no sense to say '(Mary plays tennis) $\div$ (Jim plays basketball)' or to say '$5 \;\rightarrow\; 7$'. If you did not know this you could be easily confused by the symbol, $-$, which has two different meanings. In *front of a term* it is a unary operation sign that makes a negative number out of a positive number, or makes a positive number out of a negative number. *Between two terms* it is a binary operation sign indicating subtraction.

Since a complex term is itself a term it can in turn be combined with operation signs to make more complex terms. We need some way to tell which operation sign is dominant for any complex term. What number is equal to $3+4\times5$? We can use parentheses to indicate the two possible groupings.

$$(3+4)\times5$$
$$3+(4\times5)$$

The first is then 7×5 which is 35. The second is $3+20$ which is 23.

EXERCISE 2

A. Insert parentheses to make the following statements true.

1. $2+6\times5=40$
2. $2+6\times5=32$

3. $-3^2 = -9$
4. $-3^2 = 9$
5. $12 - 3^2 = 3$
6. $12 - 3^2 = 81$
7. $24 \div 3 + 2^2 = 12$
8. $24 \div 3 + 2^2 = 3\frac{3}{7}$
9. $24 \div 3 + 2^2 = 100$
10. $24 \div 3 + 2^2 = {}^{24}\!/_{25}$

B. Give a formal proof of each of the following arguments.

1. Three plus seven is greater than two plus five.
    Anything greater than two plus five is not equal to two times three. Therefore, three plus seven is not equal to two times three.

2. Any number that is not equal to zero is greater than zero or is less than zero.
    Six divided by two is not zero and six divided by two is not less than zero.
    Therefore, six divided by two is greater than zero.

3. A number is even if and only if it is divisible by two.
    Three times five is not even but three plus five is divisible by two.
    Therefore, three times five is not divisible by two but three plus five is even.

4. For all x, x plus one is even or x is not odd. If one plus three is not even then three plus one is not even.
    Therefore, if three is odd then one plus three is even.

5. Three added to any odd number gives an even number.
    (Hint: If a number is odd then that number plus three is even.)
    Two plus three is odd.
    If the result of adding three to two plus three is even then eight is even.
    Therefore, eight is even.

We are not restricted in an argument to a single premise using a universal quantifier. The following example illustrates this point.

> For every x, if x is an even number then $x + 2$ is an even number.
> For every x, if x is an even number then x is not an odd number.

Two is an even number.

Therefore, $2+2$ is not an odd number.

On defining, $Ex \leftrightarrow x$ is an even number and $Dx \leftrightarrow x$ is an odd number, we symbolize the argument and write a derivation,

$$
\begin{array}{lll}
(1) & (\forall x)(Ex \rightarrow E(x+2)) & P \\
(2) & (\forall x)(Ex \rightarrow \neg Dx) & P \\
(3) & E2 & P \\
(4) & E2 \rightarrow E(2+2) & 2/x \quad 1 \\
(5) & E(2+2) & PP\ 4 \\
(6) & E(2+2) \rightarrow \neg D(2+2) & 2+2/x \quad 2 \\
(7) & \neg D(2+2) & PP\ 5,\ 6
\end{array}
$$

In this case a universal quantifier occurs in lines (1) and (2). In line (4) we apply US to (1), putting '2' for 'x'. In line (6) we apply US to (2), but this time the structure of the argument calls for us to put '$2+2$' for 'x'.

EXERCISE 3

A. Symbolize the following arguments using logical symbols and the standard symbols from arithmetic such as $+$, $>$, $<$, and so on, whenever appropriate.

1. For every y, if y is less than 9 then y is less than 10.
   $4+4$ is less than 9.
   Therefore, $4+4$ is less than 10.
2. For every x, if x is greater than four then x is greater than three.
   One plus one is not greater than three.
   Therefore, one plus one is not greater than four.
3. For every z, if z is equal to three plus one then z is equal to two plus two.
   Eight minus four is equal to three plus one.
   Therefore, eight minus four is equal to two plus two.
4. Every negative number is less than zero.
   Two is not less than zero.
   Therefore, two is not a negative number.
5. For every x, if $x+1=1$ then x is less than 1.
   $0+1=1$.
   Therefore, 0 is less than 1.

6. Every number divisible by two is even.
   Four is either odd or it is a number divisible by two.
   Four is not odd.
   Therefore, four is even.
7. For every y, if y is the sum of even numbers then y is an even number.
   Eight is the sum of even numbers.
   Twelve is the sum of even numbers.
   Therefore, both eight and twelve are even numbers.
8. For every x, if x is equal to ten then x is greater than eight.
   Five plus five is equal to ten or five plus three is equal to ten.
   Five plus three is not equal to ten.
   Therefore, five plus five is greater than eight.
9. For every x, it is not the case that x is both a positive number and x is a negative number.
   For every x, if x is less than 0 then x is a negative number.
   $1+1$ is a positive number.
   Therefore, $1+1$ is not less than 0.
10. For every x, if x is greater than 2 then $x+2$ is greater than 2.
    For every x, if $x+1$ is greater than 2 then $x+2$ is greater than 2.
    Either 2 is greater than 2 or $2+1$ is greater than 2.
    Therefore, $2+2$ is greater than 2.
11. All spiders are arachnids.
    All arachnids have eight legs.
    Charlotte is a spider.
    Therefore, Charlotte has eight legs.
12. No triangle congruent to ABC is equilateral.
    Only triangles congruent to ABC are congruent to DEF.
    Triangle GHI is equilateral.
    Therefore, triangle GHI is not congruent to DEF.

B.   Write a full derivation for each of the arguments in Exercise **A**.

C.   Prove the following conclusions by showing a complete derivation from the premises.

> 1. Prove: $2+0>1$
>   (1) $(\forall x)(x=2 \quad \rightarrow \quad x=1+1)$
>   (2) $(\forall x)(x=1+1 \quad \rightarrow \quad x>1)$
>   (3) $2+0=2$

2. Prove: $\neg N3$
   (1) $(\forall x)(Nx \rightarrow x<0)$
   (2) $\neg(3<0)$

3. Prove: $Fa \rightarrow La$
   (1) $(\forall x)(Fx \rightarrow \neg Px)$
   (2) $(\forall x)(Px \lor Lx)$

4. Prove: $\neg N4$
   (1) $(\forall x)(x>0 \leftrightarrow Px)$
   (2) $(\forall x)(Px \rightarrow \neg Nx)$
   (3) $4>0$

5. Prove: $2\times3\neq0$
   (1) $(\forall y)(Py \lor Ny \rightarrow y\neq0)$
   (2) $P(2\times3)$

6. Prove: $5-5=0$
   (1) $(\forall x)(\neg Px \rightarrow (\neg Nx \rightarrow x=0))$
   (2) $\neg N(5-5)$
   (3) $(\forall x)(x>0 \leftrightarrow Px)$
   (4) $5-5\not>0$

7. Prove: $3<5$
   (1) $(\forall x)(x<4 \ \& \ 4<5 \rightarrow x<5)$
   (2) $(\forall z)(-4<-z \leftrightarrow z<4)$
   (3) $4<5$
   (4) $-4<-3$

8. Prove: $Sb \ \& \ Pb \rightarrow \neg Cb$
   (1) $(\forall u)(Su \ \& \ Ru \rightarrow \neg Cu)$
   (2) $(\forall u)(Pu \rightarrow Ru)$

9. Prove: $12=4\times3$
   (1) $(\forall v)(12=v\times3 \leftrightarrow 3+1=v)$
   (2) $(\forall v)(v+1=4 \leftrightarrow 8-v=5)$
   (3) $8-3=5$

10. Prove: $3+4<3+7$
   (1) $(\forall x)(x<2+6 \rightarrow x<3+7)$
   (2) $(\forall x)(x>2+5 \lor x<2+6)$
   (3) $3+4\not>2+5$

In different premises using quantifiers we are not restricted to using the same variable every time. Suppose we put '*x*' as the only variable in the first premise. When we write another premise using a quantifier, we can use '*x*' again or we may just as well use a different variable, say '*y*'. The following example illustrates this point, using as before, '*Ex*' for '*x* is an even number', '*Ox*' for '*x* is an odd number', and '*Px*' for '*x* is a positive number'.

Prove: $E4$
(1) $(\forall x)(x>0 \rightarrow Ex \lor Ox)$
(2) $(\forall x)(Px \rightarrow x>0)$
(3) $P4$
(4) $\neg O4$

This argument can just as well be written with the second premise, expressed as

$$(\forall y)(Py \rightarrow y>0),$$

because we may use either '*x*' or '*y*' to talk about all numbers. On symbolizing the second premise with '*y*', the derivation is

(1) $(\forall x)(x>0 \rightarrow Ex \lor Ox)$		P
(2) $(\forall y)(Py \rightarrow y>0)$		P
(3) $P4$		P
(4) $\neg O4$		P
(5) $P4 \rightarrow 4>0$		$4/y$ 2
(6) $4>0$		PP 3, 5
(7) $4>0 \rightarrow E4 \lor O4$		$4/x$ 1
(8) $E4 \lor O4$		PP 6, 7
(9) $E4$		TP 4, 8

Note that we put '4' for '*x*' and later put '4' for '*y*.' We may specify any term in either universal sentence.

EXERCISE 4

Symbolize the following arguments using logical symbols and standard symbols of arithmetic. Then write a full derivation of the conclusion.

1. For all *y*, *y* is even if and only if $y+1$ is odd.
   For all *x*, if *x* equals $5+1$ then *x* is even.
   $5+1$ is not odd.
   Therefore, 5 does not equal $5+1$.

2. For all x, if $12 = x + 4$ or $x = 5 \times 3$ then x is not even.
   For every y, y is even or y is odd.
   $15 = 5 \times 3$.
   Therefore, 15 is odd.

3. For all z, if z is greater than three plus four then z is greater than zero.
   Any y is positive if and only if it is greater than zero.
   Three plus five is greater than three plus four.
   Therefore, three plus five is positive.

4. Every pupil who has done his homework understands the problem.
   Ernie is a pupil but he doesn't understand the problem.
   Therefore, Ernie hasn't done his homework.

5. Whoever was the composer of the "1812 Overture" deserved an early grave.
   Scarlatti wrote for the harpsichord.
   No one is both a true musician and deserves an early grave.
   All who wrote for the harpsichord were true musicians.
   Therefore, Scarlatti was not the composer of the "1812 Overture."

6. For all x, if the square of x is nine and x is greater than two then x is three.
   Every y would be less than four if whenever it is greater than two it equals three.
   The square of one plus two is nine.
   Therefore, one plus two is less than four.

7. Any u that equals three plus five or equals ten plus two is divisible by four.
   Any x that is divisible by four or divisible by six is even.
   Therefore, if nine minus one equals three plus five then nine minus one is even.

8. Every x divisible by twelve is divisible by four.
   Every y divisible by four is even.
   Either a z is divisible by two or it is not even.
   Fifteen is not divisible by two.
   Therefore, fifteen is not divisible by twelve.

9. Every number is either less than five or both greater than three and positive.
   If a number is greater than zero then if it is less than five it is positive.

Four is a number greater than zero.
Therefore, four is positive.

10. Every x is either greater than zero or it is not positive.
No y which multiplied by three gives minus six is greater than zero.
Therefore, if four plus five is positive then three times four-plus-five, $3 \times (4+5)$, does not equal minus six.

▶ 6.2 *Two or More Quantifiers*

We cannot do much mathematics or other systematic thinking always using just one quantifier with each sentence, for in mathematics we are always dealing with relations between two or more things. Fortunately, it is extremely simple to extend everything we have done to include sentences involving more than one universal quantifier so long as all the quantifiers are at the beginning of the sentence.

As an example, consider the argument:

For every x and y, if x is greater than y,
then it is not the case that y is greater than x.
Two is greater than one.
Therefore, it is not the case that one is greater than two.

We may symbolize this argument,

Prove: $\neg(1>2)$
(1) $(\forall x)(\forall y)(x>y \rightarrow \neg(y > x))$
(2) $2>1$

Note that we simply put two universal quantifiers, one using the variable 'x' and the other the variable 'y' at the beginning of the first premise. To each of these variables we may apply universal specification, and we replace 'x' by '2' and 'y' by '1'. A full derivation has the form:

(1) $(\forall x)(\forall y)(x>y \rightarrow \neg(y>x))$ P
(2) $2>1$ P
(3) $2>1 \rightarrow \neg(1>2)$ $2/x, 1/y$ 1
(4) $\neg(1>2)$ PP 2, 3

As a second example, consider the argument:

> For every x and y, if x is equal to y, then y is equal to x.
> One plus one is equal to two.
> Therefore, two is equal to one plus one.

In symbolizing this argument, we use the standard sign for equality, $=$.

$$\text{Prove: } 2 = 1 + 1$$
$$(1)\ (\forall x)(\forall y)(x = y\ \rightarrow\ y = x)$$
$$(2)\ 1 + 1 = 2$$

As in the first example to apply universal specification we substitute '$1 + 1$' for 'x' and '2' for 'y', in order to have the second premise serve as the antecedent of the implication. Once the substitution is made we need only apply *modus ponendo ponens* to obtain the desired conclusion. The full derivation is

(1) $(\forall x)(\forall y)(x = y\ \rightarrow\ y = x)$	P	
(2) $1 + 1 = 2$	P	
(3) $1 + 1 = 2\ \rightarrow\ 2 = 1 + 1$	$1 + 1/x, 2/y$	1
(4) $2 = 1 + 1$	PP 2, 3	

Notice that '2 is equal to $1 + 1$' and '2 is greater than 1' are atomic formulas but contain two terms. They express some mathematical relation between their terms. There are also many nonmathematical relations. 'Elizabeth II is the mother of Prince Charles' and 'Booth killed Lincoln' express relations. Such relations are two-place predicates. On defining

$$Mxy\ \leftrightarrow\ x \text{ is the mother of } y$$
$$e = \text{Elizabeth II}$$
$$c = \text{Prince Charles}$$

the first sentence can be symbolized:

$$Mec, \text{ or } eMc.$$

Similarly, the second one can be symbolized:

$$Kbl.$$

When we use equivalences like '$Mxy\ \leftrightarrow\ x$ is the mother of y' we are saying that whenever we have 'is a mother of' in English we will use 'M' in symbols and whenever we have 'M' in symbols it means 'is mother of' in English. 'x' and 'y' are used to show that 'is mother of' and '*M*' are two-place predicates requiring two terms. The terms '*e*' and '*c*' are not used in giving the translation of 'is mother of' because the translation of each term and each predicate must be given separately to keep them clear. We separately give translations for 'Elizabeth', 'Charles', and 'is the mother of', but not for 'Elizabeth is the mother of Charles' all at once.

Often it may seem better to write '*xMy*' instead of '*Mxy*'. Either one is acceptable.

Counting the number of terms is not enough to determine whether an English sentence is to be translated using one-place or two-place predicates.

For example,

> Wilbur and Orville are men

means

> Wilbur is a man and Orville is a man.

On defining,

$$Mx \quad \leftrightarrow \quad x \text{ is a man}$$
$$w = \text{Wilbur}$$
$$o = \text{Orville}$$

In symbols

$$Mw \quad \& \quad Mo.$$

Consider

> Wilbur and Orville are brothers.

This does not mean

> Wilbur is a brother and Orville is a brother

but rather,

> Wilbur is a brother of Orville.

On defining,

$$Bxy \quad \leftrightarrow \quad x \text{ is brother of } y$$
$$w = \text{Wilbur}$$
$$o = \text{Orville.}$$

In symbols

$$Bwo.$$

As another example, symbolize

If Frances is the wife of Francis then Francis is a man.

On defining,

$$Wxy \quad \leftrightarrow \quad x \text{ is the wife of } y$$
$$Mx \quad \leftrightarrow \quad x \text{ is a man}$$
$$e = \text{Frances}$$
$$i = \text{Francis.}$$

In symbols

$$Wei \quad \rightarrow \quad Mi.$$

The following example requires quantifiers.

Every man is older than any boy.

On defining,

$$Mx \quad \leftrightarrow \quad x \text{ is a man}$$
$$Bx \quad \leftrightarrow \quad x \text{ is a boy}$$
$$Oxy \quad \leftrightarrow \quad x \text{ is older than } y,$$

then the sentence is symbolized

$$(\forall x)(\forall y)(Mx \quad \& \quad By \quad \rightarrow \quad Oxy).$$

EXERCISE 5

A.   Translate the following into logical symbols.

1. Jim teases Frances.
2. Mrs. Warren visited the Huntington Library.
3. If Larry visits Harry then Mary visits Sherry.
4. All things attract each other.
5. Every eagle is larger than any hummingbird.
6. Brothers sometimes quarrel with each other.
7. Any boy who wants to play with the volley ball must first pump it up.
8. Earl helps Betty and is helped by Jennifer.
9. Birds are afraid of cats.
10. Every nation which fears another prepares to fight it.

B.   Translate the following arguments into logical symbols and give a derivation of the conclusion from the premises.

1. Everything in this lesson is part of logic.
   Any person who can figure out any part of logic is a genius.
   Karen is a person who can figure out the first derivation, and it is in this lesson.
   Therefore, Karen is a genius.
2. Mongooses can kill cobras.
   Montgomery cannot kill Charlie.
   Therefore, if Charlie is a cobra then Montgomery is not a mongoose.
3. Anyone who likes George will choose Nick for his team.
   Nick is not a friend of anyone who is a friend of Mike.
   Jay will choose no one but a friend of Ken for his team.
   Therefore, if Ken is a friend of Mike then Jay does not like George.
4. Only a fool would feed a wild bear.
   Christine fed Nicholas but she is no fool.
   Therefore, Nicholas is not a wild bear.

C. Derive the required conclusion from the given premises. Here: $Ex \leftrightarrow x$ is even, $Ox \leftrightarrow x$ is odd, $Px \leftrightarrow x$ is positive, $Nx \leftrightarrow x$ is negative, $Dx \leftrightarrow x$ is divisible by two.

1. Prove: $5+3=3+5$
   (1) $(\forall x)(\forall y)(x+y=y+x)$

2. Prove: $(4+3)+3>6+3$
   (1) $(\forall x)(\forall y)(x>y \rightarrow x+3>y+3)$
   (2) $4+3>6$

3. Prove: $3/4<1$
   (1) $(\forall w)(\forall z)(w>z \rightarrow z/w<1)$
   (2) $4>3$

4. Prove: $E(3\cdot8)$*
   (1) $(\forall x)(\forall y)(Ex \rightarrow E(x\cdot y))$
   (2) $(\forall u)(\forall v)(E(u\cdot v) \leftrightarrow E(v\cdot u))$
   (3) $E8$

5. Prove: $-4\cdot(-4)^2<-4$
   (1) $(\forall x)(\forall y)(x<-1 \quad \& \quad y>1 \rightarrow x\cdot y<x)$
   (2) $(\forall z)(z<-1 \rightarrow z^2>1)$
   (3) $-4<-1$

---

* The $\cdot$ is used frequently for multiplication in order to avoid confusing the multiplication sign with the variable x.

6. Prove: $6 \cdot 2/3 < 6$
   (1) $(\forall u)(\forall v)(u - v < 0 \quad \leftrightarrow \quad u < v)$
   (2) $(\forall w)(\forall z)(w < 1 \quad \& \quad w > 0 \quad \rightarrow \quad z \cdot w < z)$
   (3) $2/3 - 1 < 0 \quad \& \quad 2/3 > 0$

7. Prove: $E(5 + 7)$
   (1) $(\forall y)(\forall z)(Oy \quad \& \quad Oz \quad \rightarrow \quad E(y + z))$
   (2) $(\forall x)(Ox \quad \leftrightarrow \quad \neg Dx)$
   (3) $(\forall w)(Ow \quad \lor \quad Ew)$
   (4) $\neg D7 \quad \& \quad \neg E5$

8. Prove: $5 + 1/4 > 5$
   (1) $(\forall x)(\forall y)(Px \quad \& \quad Py \quad \& \quad x < 1 \quad \rightarrow \quad y + x > y)$
   (2) $(\forall x)(\forall y)((Py \quad \& \quad Px) \quad \lor \quad \neg(y + x > x \quad \lor \quad y + x > y))$
   (3) $1/4 < 1$
   (4) $1/4 + 5 > 5$

9. Prove: $P5 \quad \rightarrow \quad (P(-3) \quad \leftrightarrow \quad P(5 \cdot -3))$
   (1) $(\forall z)(\forall y)(Pz \quad \& \quad Py \quad \rightarrow \quad P(z \cdot y))$
   (2) $(\forall y)(\forall w)(Py \quad \& \quad \neg Pw \quad \rightarrow \quad \neg P(y \cdot w))$

10. Prove: $P7$
    (1) $(\forall x)(\forall y)(x > 0 \quad \& \quad y < 0 \quad \rightarrow \quad Nx/y)$
    (2) $(\forall u)(\forall v)((u < 0 \quad \rightarrow \quad Nv/u) \quad \rightarrow \quad Pv)$
    (3) $7 > 0$

There is no need to restrict ourselves to two quantifiers. We may introduce as many quantifiers as needed in order to symbolize appropriately the English sentence. The following examples illustrate this point.

> For every x, y, and z, if x is greater than y and y
> is greater than z then x is greater than z.
> Two is greater than one.
> Three is greater than two.
> Therefore, three is greater than one.

We introduce three quantifiers for the variables 'x', 'y', and 'z' to symbolize the first premise. The full argument is symbolized as follows:

Prove: $3 > 1$
   (1) $(\forall x)(\forall y)(\forall z)(x > y \quad \& \quad y > z \quad \rightarrow \quad x > z)$
   (2) $2 > 1$
   (3) $3 > 2$

In this case, we must give some consideration to exactly what substitution is to be made for 'x', 'y', and 'z' by universal specification. The point is to decide how we may use our two additional premises to form the antecedent of a conditional having the conclusion as a consequent. In the present case the important clue is that we want the variable 'y' to be replaced by the name of the number which occurs both as greater than one number and as less than another number. We look at our two other premises and see that this must be '2'. The rest of the substitution is then clear. Three is greater than two and two is greater than one so we substitute '3' for 'x' and '1' for 'z'. The complete derivation is

(1) $(\forall x)(\forall y)(\forall z)(x > y \quad \& \quad y > z \quad \rightarrow \quad x > z)$ P

(2) $2 > 1$ P

(3) $3 > 2$ P

(4) $3 > 2 \quad \& \quad 2 > 1 \quad \rightarrow \quad 3 > 1$ $3/x, 2/y, 1/z$ 1

(5) $3 > 2 \quad \& \quad 2 > 1$ A 2, 3

(6) $3 > 1$ PP 4, 5

EXERCISE 6

A. Translate the following arguments into logical symbols and give a derivation of the conclusion from the premises.

1. The sister of any boy's mother is his aunt.
    Bob is a boy and Martha is Helen's sister.
    All of Bob's aunts send him birthday presents.
    Therefore, if Helen is Bob's mother, Martha sends him birthday presents.

2. Colonels outrank sergeants and sergeants outrank privates.
    Anyone who is outranked by another must take orders from him.
    Anyone who outranks another who in turn outranks a third, outranks the third.
    Werner is a colonel, Haller is a sergeant, and Bradley is a private.
    Therefore, Bradley must take orders from Werner.

3. For every x and y if x is greater than y, it is not the case that y is greater than x.
    Therefore, it is not the case that one is greater than one.
    [Hint. Try an indirect proof.]

4. For every x and y, x is equal to or greater than y or y is equal to or greater than x.

Therefore, one is equal to or greater than one.

B.   Derive the conclusion from the given premises showing a complete proof in standard form.

1. Prove: $4 = 2 + 2$
   (1) $(\forall x)(\forall y)(\forall z)(x = y \quad \& \quad y = z \quad \rightarrow \quad x = z)$
   (2) $4 = 2^2$
   (3) $2^2 = 2 + 2$

2. Prove: $2 \neq 1$
   (1) $(\forall x)(\forall y)(\forall z)(x = y \quad \& \quad x > z \quad \rightarrow \quad y > z)$
   (2) $\neg(1 > 1)$
   (3) $2 > 1$

3. Prove: $2 = 1 + 1$
   (1) $(\forall x)(\forall y)(\forall z)[x = y + 1 \quad \vee \quad (x = y \quad \& \quad y = z + 1)]$
   (2) $2 \neq 1 \quad \vee \quad 1 \neq 0 + 1$

C.   From the following premises (1) to (4) derive each of the conclusions listed below.

(1) $(\forall x)(\forall y)(\forall z)(xQy \quad \& \quad yQz \quad \rightarrow \quad xQz)$
(2) $(\forall x)(\forall y)(xQy \quad \vee \quad yQx)$
(3) $(\forall x)(\forall y)(xIy \quad \leftrightarrow \quad xQy \quad \& \quad yQx)$
(4) $(\forall x)(\forall y)(xPy \quad \leftrightarrow \quad \neg yQx)$

The conclusions are

(a) bIb (Hint: This will need (3) and (2))
(b) $aPb \quad \rightarrow \quad \neg bPa$
(c) $aPb \quad \& \quad bQc \quad \rightarrow \quad aQc$

▶ 6.3 *Logic of Identity*

In English we often put some form of the verb 'to be' (usually 'is', 'was', 'are', or 'were') between two terms to indicate that they name or refer to the same thing. For example,

(1) Elizabeth II is the Queen of England.

This means

>   (2) 'Elizabeth II' names or refers to the same thing as 'the Queen of England'.

or

>   (3) Elizabeth II is the same as the Queen of England.

or

>   (4) Elizabeth II is identical with the Queen of England.

On defining,

$$e = \text{Elizabeth II}$$
$$q = \text{the Queen of England};$$

these can be symbolized

$$e = q$$

The equals sign, $=$, is also called the 'sign of identity' or 'identity sign'.

However, the verb 'to be' is also used in other ways in English. The following two sentences have the same general appearance as (1), but they cannot be restated like (2), (3), or (4).

>   (5) Elizabeth II is a woman.
>   (6) Women are people.

It is incorrect to restate these as

>   (7) 'Elizabeth II' refers to the same thing as 'a woman'.
>   (8) Women are identical with people.

In deciding whether a sentence can be translated using the identity sign it is necessary to decide what it means. Trying it in the forms of (2), (3), or (4) and deciding whether the result means the same thing often helps.

The thing to remember is that the identity sign belongs only between terms and they must be names for one and the same thing. No two different things are one and the same thing. Even two new pencils from the same box, although they may look so much alike you could not tell them apart, are still different pencils—they are not identical. It would be wrong to say 'first pencil $=$ second pencil'; saying they are equal or identical means they are the *same* pencil not just that they are so much alike you could not tell them apart. The $=$ sign does not go between two *things* but between two *symbols* of *expressions* and what it means is that the two expressions refer to the same *one* thing. So we can say

> George Washington = the first President of the United States.

The idea is simple. But people do get confused about it because they often use the words 'equal' or 'identical' in ways that are not strictly accurate ('These pencils are identical') rather than in the precise mathematical and logical way.

EXERCISE 7

A. Translate the following sentences using the identity sign wherever appropriate.

1. July 4th is Independence Day.
2. Sir Francis Drake was a privateer.
3. Cats are felines.
4. One fifth is twenty per cent.
5. Benjamin Franklin was a printer.
6. Franklin was the author of *Poor Richard's Almanac*.
7. Monterey is not the capital of California.
8. Not every horse is a good racer.
9. If Mike is a photographer then he is the boy I heard about.
10. If x is three then it does not equal y.

Identity plays an important role in mathematical logic. Study the following example.

Example a.

$$
\begin{array}{lll}
(1) & 2>1 & P \\
(2) & 2=1+1 & P \\
(3) & 1+1>1 & I\ 1,\ 2
\end{array}
$$

Line (3) is obtained from line (1) by replacing '2' by '1 + 1'. This is justified by the Rule for Identities because line (2) tells us '2' and '1 + 1' are names for the same thing.

As an example involving variables, consider

Example b.

$$(1) \quad (\forall x)(\forall y)(x>y \quad \rightarrow \quad x+1>y) \qquad \qquad \text{P}$$
$$(2) \quad 1>0 \qquad \qquad \text{P}$$
$$(3) \quad 2=1+1 \qquad \qquad \text{P}$$
$$(4) \quad 1>0 \quad \rightarrow \quad 1+1>0 \qquad \qquad 1/x,\,0/y \quad 1$$
$$(5) \quad 1+1>0 \qquad \qquad \text{PP 2, 4}$$
$$(6) \quad 2>0 \qquad \qquad \text{I 3, 5}$$

Note the application of the Rule for Identities to obtain (6) from (3) and (5). We abbreviate the rule as I. Note also how the Rule for Identities is used. After specifying '1' for 'x' and '0' for 'y', we obtain by *ponendo ponens*

$$1+1>0.$$

We then use immediately the rule of substitution for identities to replace '$1+1$' by '2' to obtain the conclusion

$$2>0.$$

We now give a formal statement of the Rule for Identities.

Suppose **S** *is a formula containing the term* c, *and* **S** *becomes* **R** *when one or more occurrences of* c *are replaced by* d. *Then from* **S** *and the identity* c $=$ d *we can conclude* **R**.

Notice how this rule is applied in *Examples a* and *b*. In *Example a*, line (1) '$2>1$' corresponds to **S**, line (2) '$2=1+1$' is the identity, and the conclusion line (3) '$1+1>1$' is like (1) except that in (3) we replace '2' by '$1+1$'. So (3) corresponds to **R** and follows logically from (1) and (2) by the Rule for Identities. You have been using this rule to some extent in the mathematics you have already learned. All that we have done here is to make it a formal part of our logic.

The use of the rule in a nonmathematical example involving an indirect proof is the following:

Example:

The agent who found the letter was in the apartment. Now if anyone was in the apartment he was in town. If anyone was in Mexico then he was not in town. In fact, Higgins was in Mexico. Therefore, Higgins is not the agent who found the letter.

(1)	Ac	P
(2)	$(\forall x)(Ax \rightarrow Tx)$	P
(3)	$(\forall x)(Mx \rightarrow \neg Tx)$	P
(4)	Mh	P
(5)	$Mh \rightarrow \neg Th$	h/x 3
(6)	$\neg Th$	PP 4, 5
(7)	$Ah \rightarrow Th$	h/x 2
(8)	$\neg Ah$	TT 6, 7
(9)	$h = c$	P
(10)	$\neg Ac$	I 8, 9
(11)	$Ac \; \& \; \neg Ac$	A 1, 10
(12)	$h \neq c$	RAA 9, 11

EXERCISE 8

A. Symbolize the following arguments and prove that the inference is valid by deriving the conclusions.

1. All positive numbers are greater than zero.
    Three is a positive number.
    Three is equal to two plus one.
    Therefore, two plus one is greater than zero.
2. All members of the committee live in this city.
    The president of the society is a member of the committee.
    Miss Jackson is the president of the society.
    Therefore, Miss Jackson lives in this city.
3. Edwards could have seen the murderer's car.
    Ramsey was the first defense witness.
    Either Edwards was at the party or Ramsey gave false testimony.
    In fact, no one at the party could have seen the murderer's car.
    Therefore, the first defense witness gave false testimony.
4. Samuel Clemens was a river captain.
    No river captain ever ignores any sign of danger.
    Mark Twain wrote about everything he did not ignore.
    Mark Twain was Samuel Clemens.
    Therefore, if the lights on the bridge are a sign of danger then
    Mark Twain wrote about them.

B. Derive the following conclusions from the given premises, showing a complete formal proof in standard form.

1. Derive: $a \neq b$
   (1) $(\forall x)(Tx \rightarrow Bx)$
   (2) $\neg Ba$
   (3) Tb

2. Derive: $2^2 + 1 > 2^2$
   (1) $4 = 2^2$
   (2) $4 = 4$
   (3) $(\forall x)(\forall y)(x = y \rightarrow x + 1 > y)$

3. Derive: $2 + 3 = 5$
   (1) $(\forall x)(\forall y)(x + y = y + x)$
   (2) $3 + 2 = 5$

4. Derive: $3^2 \neq 6$
   (1) $(\forall x)(x < 7 \rightarrow x < 8)$
   (2) $\neg(3^2 < 8)$
   (3) $6 < 7$

5. Derive: $\neg(3^2 = 6)$
   (1) $(\forall x)(x > 7 \rightarrow \neg(x = 6))$
   (2) $3^2 = 9$
   (3) $9 > 7$

6. Derive: $E36$
   (1) $(\forall z)(z^2 = z \cdot z)$
   (2) $(\forall x)(\forall y)(Ex \rightarrow E(x \cdot y))$
   (3) $E6$
   (4) $6^2 = 36$

7. Derive: $4 + 3 \neq 3 \cdot 2$
   (1) $(\forall x)(\forall y)(x + 3 = y + 2 \rightarrow x + 1 = y)$
   (2) $4 + 1 \neq 4$
   (3) $3 \cdot 2 = 4 + 2$

8. Derive: $3 + 2 = 5$
   (1) $(\forall x)(\forall y)(\forall z)(x - y = z \leftrightarrow y + z = x)$
   (2) $5 - 3 = 1 + 1$
   (3) $1 + 1 = 2$

9. Derive: $0 \ (25)$
   (1) $(\forall u)(\forall v)(\forall w)(u + v = u + w \rightarrow v = w)$
   (2) $4 + 5^2 = 29$

(3) $(\forall x)(\forall y)(x^2=y \quad \rightarrow \quad (Ox \quad \rightarrow \quad Oy))$

(4) $4+25=29$

(5) $0\ (5)$

10. Derive: $4>-4$

(1) $(\forall x)(\forall y)(\forall z)(x>y \quad \& \quad y>z \quad \rightarrow \quad x>z)$

(2) $4>2+1$

(3) $(\forall w)(\forall z)(Pw \quad \& \quad Nz \quad \rightarrow \quad w>z)$

(4) $P3 \quad \& \quad N(-4)$

(5) $2+1=3$

11. Derive: $3\cdot7=21$

(1) $(\forall x)(\forall y)(\forall z)(x\cdot(y+z)=(x\cdot y)+(x\cdot z))$

(2) $3\cdot5=15$

(3) $3\cdot2=6$

(4) $2+5=7$

(5) $6+15=21$

▶ 6.4 *Truths of Logic*

A truth of logic is a formula that is true independent of the truth or falsity of any particular factual premises. Tautologies are examples of truths of logic. They are always true by virtue of their form. Other examples are statements of the following form:

$$1=1$$
$$x=x$$
$$\text{Lincoln}=\text{Lincoln}$$
$$3+4=3+4$$
$$\text{the evening star}=\text{the evening star}$$

It is a logical truth that anything equals itself.

The *Rule for Logical Truths* (L) allows us to make use of logical truths in our formal proofs. It is simply stated:

A logical truth may be introduced at any point in a derivation.

It is premise-free.

So adding a logical truth in a derivation is not adding a premise, but is justified by rule L.

The need for the Rule for Logical Truths is shown in the following example.

Prove: $2+1=(1+1)+1$
(1) $2=1+1$

It certainly seems that the conclusion follows from the premise. And the Rule for Logical Truths allows the conclusion to be proved.

(1) $2=1+1$ P
(2) $2+1=2+1$ L
(3) $2+1=(1+1)+1$ I 2, 1

Line (2) is a truth of logic that was constructed by selecting the left side of the desired conclusion and stating that it is equal to itself. Also notice how the Rule for Identities was used. Here line (2) corresponds to S of the rule, the right hand '2' is c and '$1+1$' is d. So (1) corresponds to $c=d$ and R, the conclusion, is obtained by putting '$1+1$' instead of the right hand '2' in line (2).

Rule L can be of use in sentential logic too. Study this proof:

(1) P → Q   P
(2) ¬P → R   P
(3) P ∨ ¬P   L
(4) Q ∨ R   DS 3, 1, 2

On page 169 we proved that P ∨ ¬P is a truth of logic. Without rule L this argument would require a nine-line, indirect proof. We have already remarked that tautologies are truths of logic. They may be derived as premise-free conclusions, as well as constructed by use of truth tables. Consider the tautology, (P → ¬P) → ¬P

Proof.

(1)   P → ¬P   P
(2)   P   P
(3)   ¬P   PP 1, 2
(4)   P & ¬P   A 2, 3
(5)   ¬P   RAA 2, 4
(6) (P → ¬P) → ¬P   CP 1, 5

EXERCISE 9

A. Derive the conclusion from the given premises in the following:

1. Derive: $3+1=(2+1)+1$
   (1) $3=2+1$

2. Derive: $4 = 2 + 2$
   (1) $2 + 2 = 4$

3. Derive: $(2 \cdot 3) \cdot 5 = 30$
   (1) $2 \cdot 3 = 6$
   (2) $6 \cdot 5 = 30$

4. Derive: $(2 \cdot 7) \cdot 15 = 14 \cdot (3 \cdot 5)$
   (1) $2 \cdot 7 = 14$
   (2) $3 \cdot 5 = 15$

5. Derive: $2 \cdot (3 + 4) = (2 \cdot 3) + (2 \cdot 4)$
   (1) $3 + 4 = 7$
   (2) $2 \cdot 7 = 14$
   (3) $6 + 8 = 14$
   (4) $2 \cdot 3 = 6$
   (5) $2 \cdot 4 = 8$

6. Derive: $8 + (5 - 2) = (2 \cdot 3) + 5$
   (1) $(\forall w)(\forall z)(w + z = z + w)$
   (2) $3 + 8 = 11$
   (3) $5 - 2 = 3$
   (4) $2 \cdot 3 = 6$
   (5) $5 + 6 = 11$

7. Derive: $(1 + 0) + 1 = 2$
   (1) $(\forall x)(x + 0 = x)$
   (2) $1 + 1 = 2$

8. Derive: $2 + (2 + 1) = 5$
   (1) $(\forall x)(\forall y)(x + y = y + x)$
   (2) $3 = 2 + 1$
   (3) $3 + 2 = 5$

9. Derive: $2 \cdot (5 \cdot 7) = 70$
   (1) $(\forall x)(\forall y)(x \cdot y = y \cdot x)$
   (2) $5 \cdot 7 = 35$
   (3) $35 \cdot 2 = 70$

10. Derive: $13 - (1 + 2) = 2 \cdot 5 \quad \rightarrow \quad 10 = 4 + 6$
    (1) $13 - 3 = 10$
    (2) $2 \cdot (2 + 3) = 4 + 6$
    (3) $1 + 2 = 3$
    (4) $2 + 3 = 5$

B. Give derivations of the following tautologies.

1. P ∨ Q → (R & Q) ∨ (P ∨ Q)
2. P & (Q ∨ R) → (P & Q) ∨ (P & R)
3. [P → (Q → R)] → [P & Q → R]
4. P & Q → ¬P ∨ Q
5. ¬(¬P → ¬Q) → ¬P & Q
6. (P ↔ Q) ↔ (¬P ↔ ¬Q)

R E V I E W   T E S T

I. Translate the following into logical symbols.

a. Charlotte rides with Ann.
b. No man can run twenty miles an hour.
c. Beethoven is the composer of 'Fidelio'.
d. Democrats don't agree with Republicans.
e. Thomas Jefferson is the author of the Declaration of Independence.
f. The Solicitor General represents the government in cases before the Supreme Court.
g. All turtles are reptiles.
h. Audubon was an American naturalist famous for his study of birds.

II. Insert parentheses in the following to make correct identities.

a. $2 \times 5 - 3 = 7$
b. $3 \times 4 + 5 = 27$
c. $4 + 5^2 = 29$
d. $4 + 5^2 = 81$
e. $24 \div 6 - 2 = 2$
f. $24 \div 6 - 2 = 6$

III. Translate the following arguments into logical symbols. Then derive the conclusions from the premises.

a. Every member of our class is either in the play or working backstage.
Those in the play are rehearsing.
Those working backstage are painting scenery.
Therefore, if Paul is a member of our class then Paul is either rehearsing or he is painting scenery.

b. Any boy is younger than his father.
Carl is a boy who is not younger than Frank.
Anyone married to Virginia is Carl's father.
Therefore, Frank is not married to Virginia.

c. Every girl in Ron's family has been on the Honors List.
Edith is a girl in Ron's family.
The recipient of the poetry award has not been on the Honors List.
Therefore, Edith is not the recipient of the poetry award.

IV.   Derive the required conclusion.

a. Prove: $\neg Ra \ \lor \ Pb$
   (1) $(\forall x)(Rx \ \rightarrow \ Sx)$
   (2) $\neg Sa$

b. Prove: $\sqrt{25}<0 \ \lor \ \sqrt{25}>0$
   (1) $(\forall x)(x<0 \ \leftrightarrow \ Nx)$
   (2) $(\forall x)(x>0 \ \leftrightarrow \ Px)$
   (3) $P(\sqrt{25}) \ \lor \ N(\sqrt{25})$

c. Prove: $3+5>2+2$
   (1) $(\forall x)(x>5 \ \lor \ x<7)$
   (2) $3+5 \not< 7$
   (3) $(\forall y)(y>5 \ \rightarrow \ y>2+2)$

d. Prove: $Mca \ \rightarrow \ Pec$
   (1) $(\forall x)(\forall y)(\neg Mxy \ \lor \ Syx)$
   (2) Ba
   (3) $(\forall u)(\forall z)(Szc \ \rightarrow \ (Bz \ \rightarrow \ Puc))$

e. Prove: $3+4>3$
   (1) $(\forall x)(\forall y)(x>y+3 \ \rightarrow \ x>y)$
   (2) $(\forall z)(\forall u)(u-3<z \ \rightarrow \ 3+z>u)$
   (3) $(3+3)-3<4$

f. Prove: $5+2=2+(2+3)$
   (1) $(\forall x)(\forall y)(x+y=y+x)$
   (2) $2+3=5$

g. Prove: $5-4=1 \ \rightarrow \ 6\cdot(5-4)=6$
   (1) $(\forall v)(v\cdot 1=v)$

CHAPTER SEVEN

A SIMPLE MATHEMATICAL SYSTEM: AXIOMS FOR ADDITION

▶ 7.1 *Commutative Axiom*

In the present chapter we show how the logic we have learned up to this point may be applied to develop, in a logical fashion, a simple mathematical system. The numerical conclusions we derive will be familiar. The starting premises will also be familiar, although their fundamental importance may not have been obvious. What is interesting and important is that a very large number of conclusions can be derived from a very few fundamental premises.

In mathematics, premises that are used again and again because of their basic and universal character are called *axioms*. This section begins with the *Commutative Axiom* for addition. The axiom says that it does not matter in what order two numbers are added. Its form is

$$(\forall x)(\forall y)(x+y=y+x).$$

By universal specification we obtain familiar statements such as

$$0+2=2+0$$
$$1+2=2+1$$
$$3+1=1+3$$
$$(2\times3)+(1+4)=(1+4)+(2\times3)$$

With the Commutative Axiom as our single premise, it is hardly possible to derive anything interesting. But we will introduce some additional premises as *definitions*. In particular we define the familiar Arabic numerals '2', '3', '4', and '5' in terms of '1'.

DEFINITIONS:

$$2=1+1$$
$$3=2+1$$
$$4=3+1$$
$$5=4+1$$

241

Each whole number is thus obtained by adding one to the previous number. Since we will be discussing addition, let us be precise about some different ways of talking about sums. Consider

$$a + b.$$

Here it is understood that we start with a and add b. It may be called 'the sum of a and b', 'b added to a', or 'a plus b'. This variety of expressions may be used to read complex sums. For example, '$(x+y)+z$' may be read 'z added to the sum of x and y'.

The four definitions for numerals and the Commutative Axiom for addition give us a total of five premises with which to work. We can now prove some very simple things. Each conclusion we prove we shall call a *theorem*. This is the standard use of the word 'theorem' in mathematics. Each theorem we prove in this chapter or assign as an exercise is a logical consequence of the axioms and definitions introduced. In other words, each theorem or conclusion proved will follow logically from the axioms and definitions given as premises. We will use the same premises in every derivation in this chapter. Therefore, it will not be necessary to write out all the premises for each example. We will simply refer to the premises (axioms and definitions) by name each time we use one. The proof of Theorem 1 will help to clarify this slight change in style.

THEOREM 1. $3 = 1 + 2$

Proof.

(1)	$3 = 2 + 1$	Def. of 3
(2)	$2 + 1 = 1 + 2$	$2/x$, $1/y$, Comm. Axiom
(3)	$3 = 1 + 2$	I 1, 2

Notice that in the *proof* of Theorem 1, we did not list all the premises but referred to them by name. We began in line (1) with one of the premises, the definition of '3'. In line (2) we used another premise, the Commutative Axiom for addition, and applied universal specification. The justification for this step is explained by "$2/x$, $1/y$ Comm. Axiom."

The reason for the change in style is that since the same premises are used in each derivation we need not show them as lines in the derivation. Of course, not all the fundamental axioms and definitions are used in *every* argument. We shall refer by name just to the premises that are used. The abbreviation for "Commutative Axiom" will be 'Comm. Axiom', and for the definitions we shall write 'Def. of 2', 'Def. of 3', and so forth.

Throughout this chapter, the fundamental axioms and any definitions that are introduced can be used as premises. In addition, as a theorem is proved we are then permitted to use that theorem in further proofs. We can often use one "already-proved" theorem in the derivation of another. This is because we have proved that the theorem follows logically from the premises and we can use it just as we can use any line in a derivation to take a further step. When an already derived theorem is used in this way, we justify the step taken by referring to Theorem 2, Theorem 5, and so forth. Theorems will be abbreviated in the following way: Theorem 1 is abbreviated as "Th. 1," and so forth.

THEOREM 2. $4 = 1 + (1 + 2)$

Proof.

(1)	$4 = 3 + 1$	Def. of 4
(2)	$3 + 1 = 1 + 3$	$3/x$, $1/y$, Comm. Axiom
(3)	$4 = 1 + 3$	I 1, 2
(4)	$4 = 1 + (1 + 2)$	I 3, Th. 1

Note that in the proof of Theorem 2, we obtain line (4) from line (3) *and* Theorem 1 by the Rule for Identities. The parentheses around '$1 + 2$' must be added in line (4) to make clear the exact point at which we apply the substitution of identical expressions.

Frequently the Rule for Logical Truths (rule L) will be helpful, especially if the expressions on both sides of the identity sign contain more than one integer. Consider the following:

THEOREM 3. $(1 + 1) + (1 + 2) = 2 + 3$

Proof.

(1)	$(1 + 1) + (1 + 2) = (1 + 1) + (1 + 2)$	L
(2)	$2 = 1 + 1$	Def. of 2
(3)	$(1 + 1) + (1 + 2) = 2 + (1 + 2)$	I 1, 2
(4)	$1 + 2 = 2 + 1$	$1/x$, $2/y$, Comm. Axiom
(5)	$1 + 2 = 3$	I 4, Def. of 3
(6)	$(1 + 1) + (1 + 2) = 2 + 3$	I 3, 5

EXERCISE 1

A.   Prove the following theorems.

THEOREM 4. $3 = 1 + (1 + 1)$
THEOREM 5. $5 = 1 + (1 + (1 + 2))$
THEOREM 6. $4 = (1 + 2) + 1$
THEOREM 7. $5 = 1 + ((2 + 1) + 1)$
THEOREM 8. $2 + (1 + 3) = 4 + 2$
THEOREM 9. $((1 + 2) + 1) + 2 = (1 + 1) + 4$

B.   These axioms, definitions, and theorems for addition can also be used in proving conclusions derived from particular given premises.

1. Prove: $4 > 3$
    (1) $(1 + 2) + 1 > (1 + 2)$

2. Prove: $x \not> 1 + 1$
    (1) $x > 2 \quad \rightarrow \quad x = (1 + 2) + 1$
    (2) $x \neq 4$

3. Prove: $x > (1 + 2) + 1$
    (1) $x = 5$
    (2) $x > 4 \quad \& \quad x < 6 \quad \leftrightarrow \quad x = 1 + (1 + (1 + 2))$

4. Prove: $3 > 2$
    (1) $(\forall x)(x + 1 > x)$

5. Prove: $5 > 2$
    (1) $(\forall x)(\forall y)(y > 0 \quad \rightarrow \quad x + y > x)$
    (2) $3 > 0$
    (3) $3 + 2 = 4 + 1$

C.   Addition is a binary operation on numbers that is commutative. Give an example of a binary numerical operation that is *not* commutative.

D.   Give an example, other than addition, of a binary numerical operation that is commutative.

E.   Give a second example of a binary numerical operation that is not commutative.

▶ 7.2 *Associative Axiom*

The operation of addition is performed between exactly two numbers. This is what is meant by saying that addition is a binary operation. Suppose we are asked for the sum $x+y+z$. We may first add x and y and then add z to the result. This can be indicated by parentheses: $(x+y)+z$. We could also add y and z first and then add the result to x as shown in $x+(y+z)$. These two ways of associating $x+y+z$ result in the same sum. This is stated in the Associative Axiom:

$$(\forall x)(\forall y)(\forall z)((x+y)+z=x+(y+z))$$

The associative axiom is sometimes called the principle of grouping for addition, meaning that it does not matter in what way numbers are grouped together to be added.

As an example, we may perform the following universal specification.

$$(3+5)+7=3+(5+7) \qquad\qquad 3/x,\ 5/y,\ 7/z.\ \text{Assoc. Axiom}$$

First, by adding inside parentheses, we have

$$8+7=3+12$$

or

$$15=15.$$

The associativity of addition is so familiar to us that it is difficult to appreciate its importance or to imagine that it could be otherwise. But subtraction is an example of a non-associative binary operation. Consider, for example,

$$8-5-2.$$

This is an ambiguous expression because

$$(8-5)-2 \text{ is } 3-2 \text{ or } 1, \text{ while } 8-(5-2) \text{ is } 8-3 \text{ or } 5.$$

In other words,

$$(8-5)-2 \neq 8-(5-2).$$

We can use the Associative Axiom to prove some theorems that express truths of arithmetic that are very familiar to you. The interesting thing is to show exactly how they follow as logical consequences of this axiom and our four definitions.

THEOREM 10. $4 = 2 + 2$

Proof.

(1) $4 = 3 + 1$	Def. of 4
(2) $4 = (2 + 1) + 1$	I 1, Def. of 3
(3) $(2 + 1) + 1 = 2 + (1 + 1)$	$2/x$, $1/y$, $1/z$, Assoc. Axiom
(4) $4 = 2 + (1 + 1)$	I 2, 3
(5) $4 = 2 + 2$	I 4, Def. of 2

Note that the Associative Axiom, abbreviated 'Assoc. Axiom', enters at exactly one point in the proof, namely, in justifying the introduction of line (3). Line (3) is obtained from the axiom by specifying '2' for 'x', '1' for 'y', and '1' for 'z'. The repeated use of rule of inference I governing identities shows that it is obviously an important general rule of inference for the kind of theorems we are now proving.

We have said that when axioms, definitions, or previous theorems are used they do not need to be written out in the proof, that they need only be referred to. This means that the proof of Theorem 10 could begin with line (2) above. This would be justified by rule I applied to the definition of 4 and definition of 3 as shown below. Either proof is correct.

(1) $4 = (2 + 1) + 1$	I Def. of 4, Def. of 3
(2) $(2 + 1) + 1 = 2 + (1 + 1)$	$2/x$, $1/y$, $1/z$, Assoc. Axiom
(3) $4 = 2 + (1 + 1)$	I 1, 2
(4) $4 = 2 + 2$	I 3, Def. of 2

EXERCISE 2

A.   Give an example, other than addition, of an arithmetical operation that is associative.

B.   Give an example, other than subtraction, of a binary arithmetical operation that is not associative.

C.   Prove the following theorems.

THEOREM 11. $5 = 3 + 2$
THEOREM 12. $3 = (1 + 1) + 1$
THEOREM 13. $4 = ((1 + 1) + 1) + 1$
THEOREM 14. $4 = (1 + (1 + 1)) + 1$
THEOREM 15. $5 = 2 + (1 + 2)$
THEOREM 16. $2 + 5 = 3 + 4$
THEOREM 17. $5 + 3 = (4 + 2) + 2$

Notice that we now have two types of rules of derivation. (1) There are the sentential rules: CP, RAA, and those on pages 108 and 109. These depend on the truth-functional properties of the connectives. (2) There are the rules governing the substitution of terms: Universal Specification and the Rule for Identities. These apply only in predicate logic. In this chapter we have been using these rules together with the axioms of commutativity and associativity and the definitions of '2', '3', '4', and '5' to develop the theory of arithmetic. The theorems in this chapter make no use of the sentential rules. But in problems 2, 3, and 5 of Exercise 1**B**, both types of rules are used. The rules governing terms are used to get different atomic formulas into identical form so that the sentential rules can be applied. For example, in problem 3 it is necessary to prove that '$x = 1 + (1 + (1 + 2))$' is equivalent to '$x = 5$'.

In some derivations, we shall want to specify variables for variables. For example, the identity

$$(x + z) + (3 + z) = x + (z + (3 + z))$$

can be derived from the associativity axiom by universal specification putting x/x, z/y, and $(3 + z)/z$. Specification of terms containing variables is permitted with one exception. Do not substitute a term containing a variable which then becomes captured by a quantifier remaining in the formula. For example, universal specification cannot properly be applied to '$(\forall x)(\forall y)(Pxy)$' putting '$y$' for '$x$' to get '$(\forall y)(Pyy)$' because the '$y$' that was substituted for 'x' becomes captured by the quantifier '$(\forall y)$'. This can be avoided by specifying for every quantifier in one step, as above, or by specifying completely new variables such as u/x, v/y, $(3 + w)/z$ in the associativity axiom, for example. This would give

$$(u + v) + (3 + w) = u + (v + (3 + w)).$$

EXERCISE 3

Give formal proofs of the following arguments.

1. Prove: $(1+2)+3>5$
   (1) $(\forall x)(x+1>x)$

2. Prove: $y=(3+z)+w$
   (1) $y=3+(z+w)$ $\vee$ $(z+w)+2\neq 2+(w+z)$

3. Prove: $(1+2)+(2+4)>(2+1)+2$
   (1) $4>0$
   (2) $(\forall x)(\forall y)(y>0 \ \rightarrow \ x+y>x)$

4. Prove: $x\neq 2+2$ & $x=1+((2+1)+1)$
   (1) $x=4 \ \rightarrow \ x+y=1+(1+(1+2))$
   (2) $x=(2+y)+1 \ \rightarrow \ x=2+(1+2)$
   (3) $\neg(x+y=2+3 \ \vee \ x\neq 3+y)$

5. Prove: $a>y+2 \ \vee \ a<y+4$
   (1) $(\forall x)(x\neq 2+3 \ \leftrightarrow \ x>(1+2)+2 \ \vee \ x<5)$
   (2) $a>3+2 \ \rightarrow \ a>(y+1)+1$
   (3) $a<1+((2+1)+1) \ \rightarrow \ a<(y+3)+1$
   (4) $a\neq 5$

6. Prove: $x<4$
   (1) $x=1+2 \ \rightarrow \ x<(1+(1+1))+1$
   (2) $x=2 \ \rightarrow \ x<1+(2+1)$
   (3) $(x^2-5x)+(2+4)=0$
   (4) $(x^2-5x)+(3+3)=0 \ \rightarrow \ x=3 \ \vee \ x=1+1$

7. Prove: $x=5+3 \ \rightarrow \ x>4$
   (1) $x=(4+2)+2 \ \rightarrow \ x>4+2$
   (2) $x<3+2 \ \rightarrow \ x\not> 1+5$
   (3) $x>3+1 \ \vee \ x<5$

When the same term appears more than once in a complex term there may be a choice as to whether to use the commutative axiom or the associative axiom. For example, note these two derivations of $(2+1)+2=2+(1+2)$.

(1) $2+(1+2)=(1+2)+2$ $2/x$, $(1+2)/y$, Comm. Axiom
(2) $1+2=2+1$ $1/x$, $2/y$, Comm. Axiom
(3) $2+(1+2)=(2+1)+2$ I 1, 2

We could use the associativity axiom to obtain the formula in just one step.

(1) $(2+1)+2=2+(1+2)$ $2/x$, $1/y$, $2/z$, Assoc. Axiom

When more than three terms appear in a complex term it may be difficult to select the specification of the associative axiom that will produce the desired result. Suppose we wish to prove the theorem, $(2+3)+(4+5)=2+(3+(4+5))$. Let us try to fit the left side, $(2+3)+(4+5)$, into the form $(x+y)+z$ or the form $x+(y+z)$. If we put one above the other and add parentheses to the axiom it is easier to see the specification required. For example,

$$(2+3)+(4+5) \quad \text{or} \quad (2+3)+(4+5),$$
$$(x+y)+(\ z\ ) \qquad\qquad (\ x\ )+(y+z)$$

The first suggests the universal specifications of $2/x$, $3/y$, $(4+5)/z$. The second suggests $(2+3)/x$, $4/y$, $5/z$. Applying the second specification, we have

$$((2+3)+4)+5=(2+3)+(4+5).$$

Although this makes the left side of the theorem appear as the right side of the specification it fails to produce the desired result on the other side. To see if it is possible in one specification to get both left and right sides of the theorem, repeat the procedure done with the left side by putting the right side of the theorem above each side of the axiom and adding parentheses to the axiom to make it fit.

$$2 \quad + \quad (3+(4+5)) \quad \text{or} \quad 2+(3+(4+5))$$
$$(x+y) \quad + \quad (\quad z\quad) \qquad\qquad x+(y+(\ z\ ))$$

The first of these suggests $3+(4+5)/z$. This is a possible specification. But it also suggests $2/x+y$. That is impossible. The second of these, however, suggests the universal specification $2/x$, $3/y$, $(4+5)/z$. This is precisely the same as the first specification suggested for the left side of the axiom so this specification will give the theorem in one step.

(1) $(2+3)+(4+5)=2+(3+(4+5))$ $2/x$, $3/y$, $(4+5)/z$,

                                                 Assoc. Axiom

Finally, we remark that the right and left sides of an identity can be interchanged by a simple derivation using rules L and I.

$$(1) \quad a=b \qquad P$$
$$(2) \quad b=b \qquad L$$
$$(3) \quad b=a \qquad I\ 1,\ 2$$

EXERCISE 4

Prove the following theorems of addition.

THEOREM 18. $5+(6+(4+3))=(5+6)+(4+3)$
THEOREM 19. $2+(3+1)=3+(2+1)$
THEOREM 20. $((2+5)+1)+7=1+(2+(5+7))$
(This requires repeated use of both Commutative and Associative Axioms.)

Sometimes it is useful to begin a proof by introducing an additional premise to be used in a conditional proof or an indirect proof. For example,

THEOREM. $2/3=5/7.5 \ \longrightarrow \ 5/7.5=2/3$

Proof.

$$(1) \qquad 2/3=5/7.5 \qquad\qquad P$$
$$(2) \qquad 5/7.5=5/7.5 \qquad\qquad L$$
$$(3) \qquad 5/7.5=2/3 \qquad\qquad I\ 1,\ 2$$
$$(4) \ \ 2/3=5/7.5 \ \longrightarrow \ 5/7.5=2/3 \qquad CP\ 1,\ 3$$

This is a truth of logic since the conclusion does not depend on any premise. It may also be called a theorem of logic. Tautologies are also truths of logic. They can be derived as premise-free conclusions. You derived several in Exercise 9 of the previous chapter.

EXERCISE 5

A.   Give proofs of the following theorems of logic.

1. $(\forall x)(Fx) \ \longrightarrow \ Fa$
2. $P \ \longrightarrow \ Q \ \lor \ P$

3. $a = b$ & $Ga \rightarrow Gb$

4. $P \lor \neg P$

5. $3 = 2 + 1$ & $b = 3 \rightarrow 2 + 1 = b$

B. Give proofs of the following theorems of arithmetic addition. This means that the axioms, definitions, and previous theorems of addition may be used in the proofs.

THEOREM 21. $(\forall x)(x + 3 > x) \rightarrow (1 + (2 + 3)) + 4 > 2 + 5$

THEOREM 22. $(3 + 1) + (2 + 4) = 5 + 5$

Occasionally in complex problems it is difficult to recognize the sentential form of the argument because atomic formulas that are equivalent may look different. At the same time it is difficult to tell which atomic formulas should be proved equivalent if you have not recognized the form of the sentential argument. In such cases it may help to recognize the sentential part of the argument if you apply your knowledge of arithmetic to simplify the atomic formulas. *This is not a logical step and is not written as part of the proof.* For example, carefully read the following:

> Prove: $x > 2 + (2 + 1)$
>
> (1) $x = (1 + 2) + (2 + 2) \rightarrow x > 3 + 2$
>
> (2) $x > (3 + 1) + 2 \rightarrow x > (2 + 1) + 2$
>
> (3) $x = 1 + (5 + 1) \lor x > 5 + 1$

None of the atomic formulas here are the same. If we added the complex terms, we would have

> Prove: $x > 5$
>
> (1) $x = 7 \rightarrow x > 5$
>
> (2) $x > 6 \rightarrow x > 5$
>
> (3) $x = 7 \lor x > 6$

In this form it is much easier to recognize that the sentential parts of the derivation will apply DS and DP. But it is necessary to use the rules governing terms to show (for the 7's) that $(1 + 2) + (2 + 2) = 1 + (5 + 1)$ and (for the 6's) that $(3 + 1) + 2 = 5 + 1$ and (for the 5's) that $(3 + 2) = (2 + 1) + 2$, and $(2 + 1) + 2 = 2 + (2 + 1)$. These are four theorems that must be proved to make the sentential derivation possible. All but the last identity must be proved before DS can be used.

A formal proof of the argument follows:

(1) $x = (1+2) + (2+2)$ $\rightarrow$ $x > 3+2$ P

(2) $x > (3+1) + 2$ $\rightarrow$ $x > (2+1) + 2$ P

(3) $x = 1 + (5+1)$ $\vee$ $x > 5+1$ P

(4) $(1+2) + (2+2) = (1+2) + (2+2)$ L

(5) $(1+2) + (2+2) = (1+2) + 4$ I 4, Th. 10

(6) $(1+2) + (2+2) = (1 + (1+1)) + 4$ I 5, Def. of 2

(7) $(1 + (1+1)) + 4 = 1 + ((1+1) + 4)$ $1/x,\ 1 + 1/y,\ 4/z,$
        Assoc. Axiom

(8) $(1+2) + (2+2) = 1 + ((1+1) + 4)$ I 6, 7

(9) $(1+1) + 4 = 4 + (1+1)$ $1 + 1/x,\ 4/y,$ Comm.
        Axiom

(10) $(1+2) + (2+2) = 1 + (4 + (1+1))$ I 8, 9

(11) $(4+1) + 1 = 4 + (1+1)$ $4/x,\ 1/y,\ 1/z,$ Assoc.
        Axiom

(12) $(1+2) + (2+2) = 1 + ((4+1) + 1)$ I 10, 11

(13) $(1+2) + (2+2) = 1 + (5+1)$ I 12, Def. of 5

(14) $(3+1) + 2 = 2 + (3+1)$ $(3+1)/x,\ 2/y,$ Comm.
        Axiom

(15) $(2+3) + 1 = 2 + (3+1)$ $2/x,\ 3/y,\ 1/z,$ Assoc.
        Axiom

(16) $(3+1) + 2 = (2+3) + 1$ I 14, 15

(17) $2+3 = 3+2$ $2/x,\ 3/y,$ Comm.
        Axiom

(18) $(3+1) + 2 = (3+2) + 1$ I 16, 17

(19) $(3+1) + 2 = 5+1$ I 18, Th. 11

(20) $3+2 = 3+2$ L

(21) $3+2 = (2+1) + 2$ I 20, Def. of 3

(22) $(2+1) + 2 = 2 + (2+1)$ $(2+1)/x,\ 2/y,$ Comm.
        Axiom

(23) $x = (1+2) + (2+2)$ $\vee$ $x > (3+1) + 2$ I 3; 13, 19

(24) $x > 3+2$ $\vee$ $x > (2+1) + 2$ DS 23, 1, 2

(25) $x > (2+1) + 2$ $\vee$ $x > (2+1) + 2$ I 24, 21

(26) $x > (2+1) + 2$ DP 25

(27) $x > 2 + (2+1)$ I 26, 22

The four numerical theorems were derived in lines (4) to (13), (14) to (19), (20) to (21), and (22). Since previously proved theorems may be used in proofs, we could have derived them in four separate proofs. If this had been done in the proof of our argument, it would have required only lines (1), (2), and (3) to state the premises and lines (23) to (27) to apply the theorems and carry out the sentential derivation.

The strategy is the following. First, determine which of the atomic formulas are equivalent so as to recognize the sentential form of the argument and decide what identities are needed. Second, prove the required identities—lines (4) to (22). Third, apply the identities to the premises to obtain formulas in which equivalent atomic formulas appear in identical form—lines (23), (25), and (27). Fourth, carry out the sentential derivation—lines (24) and (26).

If the required identities are proved separately, the derivation would be broken up into parts. We call each of the parts asserting an identity a *lemma*. As mathematicians sometimes jokingly say, a lemma is a "tiny theorem." More seriously, a lemma is something that is proved mainly for the sake of helping to prove something else. We emphasize that from a logical standpoint a lemma has exactly the same status as a theorem.

To prove, then, that $x > 2 + (2 + 1)$ from the three given premises, we first prove three lemmas that do not depend on any of the three premises but only on our axioms and definitions. Lemma A is line (13) in the proof just shown, Lemma B is line (19) and Lemma C is line (21).

LEMMA A. $(1 + 2) + (2 + 2) = 1 + (5 + 1)$

Proof.

(1) $(1 + 2) + (2 + 2) = (1 + 2) + (2 + 2)$ L

(2) $(1 + 2) + (2 + 2) = (1 + 2) + 4$ I 1, Th. 10

(3) $(1 + 2) + (2 + 2) = (1 + (1 + 1)) + 4$ I 2, Def. of 2

(4) $(1 + (1 + 1)) + 4 = 1 + ((1 + 1) + 4)$ $1/x,\ 1 + 1/y,\ 4/z$,
          Assoc. Axiom

(5) $(1 + 2) + (2 + 2) = 1 + ((1 + 1) + 4)$ I 3, 4

(6) $(1 + 1) + 4 = 4 + (1 + 1)$ $1 + 1/x,\ 4/y$, Comm.
          Axiom

(7) $(1 + 2) + (2 + 2) = 1 + (4 + (1 + 1))$ I 5, 6

(8) $(4 + 1) + 1 = 4 + (1 + 1)$ $4/x,\ 1/y,\ 1/z$, Assoc.
          Axiom

(9) $(1 + 2) + (2 + 2) = 1 + ((4 + 1) + 1)$ I 7, 8

(10) $(1 + 2) + (2 + 2) = 1 + (5 + 1)$ I 9, Def. of 5

LEMMA B. $(3+1)+2=5+1$

Proof.

(1) $(3+1)+2=2+(3+1)$	$(3+1)/x$, $2/y$, Comm. Axiom
(2) $(2+3)+1=2+(3+1)$	$2/x$, $3/y$, $1/z$, Assoc. Axiom
(3) $(3+1)+2=(2+3)+1$	I 1, 2
(4) $2+3=3+2$	$2/x$, $3/y$, Comm. Axiom
(5) $(3+1)+2=(3+2)+1$	I 3, 4
(6) $(3+1)+2=5+1$	I 5, Th. 11

LEMMA C. $3+2=(2+1)+2$

Proof.

(1) $3+2=3+2$	L
(2) $3+2=(2+1)+2$	I 1, Def. of 3

We do not state line (22) as a separate lemma because its proof is so simple it may be included in the main derivation, which now follows:

(1) $x=(1+2)+(2+2)$ $\rightarrow$ $x>3+2$	P
(2) $x>(3+1)+2$ $\rightarrow$ $x>(2+1)+2$	P
(3) $x=1+(5+1)$ $\lor$ $x>5+1$	P
(4) $x=(1+2)+(2+2)$ $\lor$ $x>(3+1)+2$	I 3, Lemmas A and B
(5) $x>3+2$ $\lor$ $x>(2+1)+2$	DS 1, 2, 4
(6) $x>(2+1)+2$ $\lor$ $x>(2+1)+2$	I 5, Lemma C
(7) $x>(2+1)+2$	DP 6
(8) $(2+1)+2=2+(2+1)$	$2+1/x$, $2/y$, Comm. Axiom
(9) $x>2+(2+1)$	I 7, 8

Notice that by first proving the three lemmas the main derivation is now only a third of its former length. From this point on you should introduce such lemmas when it will help to simplify or shorten proofs.

EXERCISE 6

Give a formal proof of each of the following arguments.
(Hint: It is convenient to prove lemmas as separate proofs.)

1. Prove: $x>4+4$ $\rightarrow$ $x>2+(1+2)$
    (1) $x>2+(3+3)$ $\rightarrow$ $x>5$
    (2) $x>(4+2)+2$ $\rightarrow$ $x>5+3$

2. Prove: $y=4$

    (1) $x=((1+2)+3)+4$ $\leftrightarrow$ $y=5+(3+5)$

    (2) $x=5+(1+4)$ $\vee$ $y=2+2$

    (3) $\neg[x=(4+3)+3$ $\vee$ $y=2+((3+3)+5)]$

▶ 7.3 *Axiom for Zero*

We now introduce the Axiom for Zero which we shall abbreviate 'Zero Axiom'.

$$(\forall x)(x+0=x)$$

The axiom says something we all know. If we add zero to any number x the result is the same number x.

    Theorems that depend on this axiom alone are not very interesting. We give one and state some others in the exercises.

THEOREM 23. $(1+0)+0=1$

    Proof.

(1)	$(1+0)+0=(1+0)+0$	L
(2)	$1+0=1$	$1/x$, Zero Axiom
(3)	$(1+0)+0=1+0$	I 1, 2
(4)	$(1+0)+0=1$	I 2, 3

Note that in line (1) a truth of logic has been introduced, as indicated by the 'L' at the right. In this case it is just an instance of the fact that $t=t$ for any term t. Note also that after obtaining line (2) by specifying '1' for 'x' in the Axiom for Zero, we use this line twice to obtain the desired conclusion.

    We obtain a variant of this theorem by also using the Associative Axiom.

THEOREM 24. $1+(0+0)=1$

    Proof.

(1)	$(1+0)+0=1+(0+0)$	$1/x$, $0/y$, $0/z$, Assoc. Axiom
(2)	$1+(0+0)=1+(0+0)$	L
(3)	$1+(0+0)=(1+0)+0$	I 1, 2
(4)	$1+(0+0)=1$	I 3, Th. 23

Using the Commutative Axiom, we can prove:

THEOREM 25. $2 = 1 + (0 + 1)$

Proof.

(1) $2 = 1 + 1$ Def. of 2
(2) $1 + 0 = 0 + 1$ $1/x$, $0/y$, Comm. Axiom
(3) $1 + 0 = 1$ $1/x$, Zero Axiom
(4) $0 + 1 = 1$ I 2, 3
(5) $2 = 1 + (0 + 1)$ I 1, 4

The theorems so far proved are very familiar truths of arithmetic. This fact does not mean their proofs are trivial. Our purpose is to show that innumerable facts of arithmetic can be logically derived from just a few fundamental axioms. The definitions simply allow shorter ways of representing things. Without the definitions of '2', '3', '4', and '5' the number five would have to be represented by:

$1 + (1 + (1 + (1 + 1)))$, or $1 + ((1 + 1) + (1 + 1))$ or $(1 + (1 + (1 + 1))) + 1$, and so on.

Not every familiar statement of arithmetic can be proved from our definitions and the three axioms introduced so far. For example, we cannot prove that $1 \neq 0$. It turns out that it is necessary to add this assertion as an axiom.

Just as not all truths of arithmetic can be proved, so also not all numerals or predicates can be defined. It is necessary to begin somewhere with undefined terms, like '1' and '0', as well as unproved statements like the axioms. This is the way we distinguish between what is fundamental and what is derived or defined. The whole collection of undefined terms, axioms, definitions, and theorems is called a *theory*.

EXERCISE 7

A.   Prove the following theorems.

THEOREM 26. $(2 + 0) + 0 = 2$
THEOREM 27. $3 + (0 + 0) = 3$
THEOREM 28. $4 = 2 + (0 + 2)$
THEOREM 29. $5 = 2 + (0 + (3 + 0))$
THEOREM 30. $4 + (0 + 3) = ((0 + 5) + 0) + 2$

B. Give formal proofs of the following arguments within the theory of arithmetic so far developed. This means you may use the axioms and definitions so far given, and the theorems so far proved, as well as the premises of each argument.

1. Prove: $1 > 0$
   (1) $(\forall x)(x + 1 > x)$

2. Prove: $x = y + 4$
   (1) $(x + 3) + 0 = (y + 2) + (3 + 2) \quad \leftrightarrow \quad x = (y + 0) + 4$
   (2) $y \neq 7$
   (3) $1 + (2 + x) = 4 + (3 + y) \quad \vee \quad y = 7$

3. Prove: $x + 2 = y + (4 + 2) \quad \& \quad y = 3$
   (1) $x + 2 \neq y + (2 + (1 + 3)) \quad \rightarrow \quad (x + 0) + 1 \neq (2 + (2 + y)) + 1$
   (2) $x + 1 = (3 + 0) + (y + 2) \quad \& \quad y + 0 = (1 + 0) + 2$

C. The Axiom for Zero is sometimes described as asserting that zero is a right-hand identity element for addition because adding zero to a number gives an identical number as sum. Which of the following statements are true?

1. Zero is an identity element for multiplication:
   $$(\forall x)(x \cdot 0 = x).$$
2. Zero is a right-hand identity element for subtraction:
   $$(\forall x)(x - 0 = x).$$
3. Zero is a left-hand identity element for subtraction:
   $$(\forall x)(0 - x = x).$$
4. Zero is a left-hand identity element for division:
   $$(\forall x)(0 \div x = x).$$
5. One is an identity element for addition:
   $$(\forall x)(x + 1 = x).$$
6. One is an identity element for multiplication:
   $$(\forall x)(x \cdot 1 = x).$$
7. One is a right-hand identity element for subtraction:
   $$(\forall x)(x - 1 = x).$$
8. One is a right-hand identity element for division:
   $$(\forall x)(x \div 1 = x).$$
9. One is a left-hand identity element for division:
   $$(\forall x)(1 \div x = x).$$

▶ 7.4 *Axiom for Negative Numbers*

In the last section we introduced zero. We now introduce negative numbers by stating the fundamental axiom for the operation $-x$ of obtaining the negative of x. We abbreviate this as 'Neg. Axiom'.

$$(\forall x)(x + (-x) = 0).$$

Examples of negative numbers are -1, -2, -3, and so on. In the following sequence of numbers

$$\ldots, 6, 5, 4, 3, 2, 1, 0$$

each number is less than the number represented on its left. Negative numbers allow this sequence to go on indefinitely in both directions.

$$\ldots, 8, 7, 6, 5, 4, 3, 2, 1, 0, -1, -2, -3, -4, -5, \ldots.$$

This entire set of numbers is called the integers. Thus the set of integers includes the positive integers, zero, and the negative integers. Each number in the sequence is less than any number represented to its left. So $-4 < -3$, $-5 < 3$, $-3 < 5$. Also for each number there is another number which is as much less than zero as the first number is greater than zero, or vice versa. The operation of obtaining the negative of x picks out that number. If $x = 5$ then $-x = -5$; and if $x = -3$ then $-x = 3$. In general

$$(\forall x)(\forall y)(-x = y \quad \rightarrow \quad -y = x).$$

The operation $-x$ of obtaining the negative of x takes any term and generates a different term. Therefore, '$-x$' is a term. The parentheses around '$-x$' in the axiom may be omitted unless it is necessary to show the intended grouping. Formulas are usually easier to read if unnecessary parentheses are omitted so we shall omit them whenever possible.

The following theorem, although simple in statement, is difficult to prove. In particular, its proof depends on all four of the axioms we have introduced in this chapter.

THEOREM 31. $1 + (2 + (-1)) = 2$

Proof.

(1) $1+(2+-1)=1+(2+-1)$	L
(2) $(1+2)+-1=1+(2+-1)$	$1/x,\ 2/y,\ -1/z$, Assoc. Axiom
(3) $1+(2+-1)=(1+2)+-1$	I 1, 2
(4) $1+2=2+1$	$1/x,\ 2/y$, Comm. Axiom
(5) $1+(2+-1)=(2+1)+-1$	I 3, 4
(6) $(2+1)+-1=2+(1+-1)$	$2/x,\ 1/y,\ -1/z$, Assoc. Axiom
(7) $1+(2+-1)=2+(1+-1)$	I 5, 6
(8) $1+-1=0$	$1/x$, Neg. Axiom
(9) $1+(2+-1)=2+0$	I 7, 8
(10) $2+0=2$	$2/x$, Zero Axiom
(11) $1+(2+-1)=2$	I 9, 10

By making our rule of inference governing identities more liberal it is possible to shorten the proof and more importantly to make its essential structure clearer. First we could eliminate line (1) if we could interchange automatically the left and right sides of an identity, in this case the Associative Axiom. If we simply specify $1/x$, $2/y$, and $-1/z$ in the axiom, we obtain line (2)

$$(1+2)+-1=1+(2+-1).$$

But we really want to begin with '$1+(2+-1)$' on the left side, for this is the left side of the identity we are trying to prove. We will henceforth use the more liberal rule that the two sides of an identity may be interchanged automatically, without citing the rule. We may at once replace the first three lines by

(1) $1+(2+-1)=(1+2)+-1$	$1/x,\ 2/y,\ -1/z$, Assoc. Axiom

In similar fashion we will eliminate the frequent citing of rule I, leaving its use understood and proceed in the following manner:

(1) $1+(2+-1)=(1+2)+-1$	$1/x,\ 2/y,\ -1/z$, Assoc. Axiom
(2) $\quad\quad\quad=(2+1)+-1$	$2/x,\ 1/y$, Comm. Axiom
(3) $\quad\quad\quad=2+(1+-1)$	$2/x,\ 1/y,\ -1/z$, Assoc. Axiom
(4) $\quad\quad\quad=2+0$	$1/x$, Neg. Axiom
(5) $\quad\quad\quad=2$	$2/x$, Zero Axiom

As long as the left side of the identity remains the same in every line, we will not write it each time. The final line, line (5), thus is

$$1+(2+-1)=2.$$

The proof has simply been reduced to a succession of identities, showing in each step what specification was made in applying the axiom.
Note that in lines (1), (3), and (5) the whole identity was obtained by
specifying in an axiom. In lines (2) and (4) only part of the identity
was considered. Thus for line (2), we used the specification $2/x$, $1/y$
on the Commutative Axiom

$$2+1=1+2$$

to replace '$2+1$' by '$1+2$' and obtain line (2) from (1).

   This new style of proof may also be illustrated by proving another
theorem similar to the last one.

   THEOREM 32. $-2+(1+2)=1$

   Proof.
(1) $-2+(1+2) = -2+(2+1)$ $1/x$, $2/y$, Comm. Axiom
(2) $= (-2+2)+1$ $-2/x$, $2/y$, $1/z$, Assoc. Axiom
(3) $= (2+ -2)+1$ $-2/x$, $2/y$, Comm. Axiom
(4) $= 0+1$ $-2/x$, Neg. Axiom
(5) $= 1+0$ $0/x$, $1/y$, Comm. Axiom
(6) $= 1$ $1/x$, Zero Axiom

From now on we shall use this more liberal rule governing identities
and as in the above proof we shall exclude references to rule I itself.

EXERCISE 8

A.   Prove the following theorems.

   THEOREM 33. $2+(-1)=1$
   THEOREM 34. $-1+3=2$
   THEOREM 35. $-5+1=-4$ [Hint. Proof involves adding 0 by
      the Zero Axiom and then replacing 0 by using the Negative
      Axiom.]
   THEOREM 36. $2+(1+ -2)=1$
   THEOREM 37. $-4+(3+4)=3$
   THEOREM 38. $3+(-5+ -3)=-5$
   THEOREM 39. $(2+(1+ -1))+ -2=0$
   THEOREM 40. $1+ -5=-((1+2)+1)$
   THEOREM 41. $-2+5=1+2$
   THEOREM 42. $-(2+(0+2))+(2+5)=(1+0)+(4+(2+ -4))$

B. Give proofs of the following arguments within the theory of arithmetic so far developed.

1. Prove: $(7+0)+(x+-7)=(-4+x)+(2+(0+2))$
2. Prove: $(1+0)+(1+x)=(5-3)+(2+y)$
   (1) $x=y+2 \;\leftrightarrow\; -1+(3+x)=0+(y+4)$
   (2) $x=(y+4)+-2$
3. Prove: $x^2\neq4+5 \;\rightarrow\; x=3+-1$
   (1) $(x^2+-x)+-6=-3+(-2+5)$
   (2) $x=-(2+1) \;\rightarrow\; x^2=(5+1)+(4+-1)$
   (3) $(x^2+-x)+-6=0 \;\rightarrow\; \neg(x\neq-3 \;\;\&\;\; x\neq1+1)$

C. The identity element for addition is 0 because adding 0 to a number gives identically that number. The Zero Axiom expresses this.

1. What is the identity element for multiplication?
2. Write out the corresponding axiom.

The number $-x$ is called the inverse of x with respect to addition because it has exactly the opposite effect from adding x, or stated in another way, it is called the inverse of x because $-x$ added to x gives the identity element for addition; that is, $x+(-x)=0$.

3. What is the inverse of x with respect to multiplication?
4. Write out the corresponding axiom.

D. Place the following numbers in a decreasing sequence starting with the largest.

$$2, 3, -5, 0, 4, -7, -9, 6, -15, -4$$

E. For each of the following give the inverse with respect to addition.

1. 5	6. -3
2. 2	7. -9
3. 4	8. 0
4. -6	9. -7
5. 8	10. 1

F. For each of the numbers of **E** above give the inverse with respect to multiplication.

R E V I E W   T E S T

I. State the following axioms in symbolized form.

    a. The Commutative Axiom for addition

    b. The Associative Axiom for addition

    c. The Axiom for Zero

    d. The Axiom for Negative Numbers

II. Put a 'T' for true statements and an 'F' for false statements.

    a. $(\forall x)(x-0=x)$

    b. $(\forall x)(x\cdot 0=x)$

    c. $(\forall x)(0-x=x)$

    d. $(\forall x)(x+1=x)$

    e. $(\forall x)(x\cdot 1=x)$

    f. $(\forall x)(x\div 1=x)$

III. Prove the following using the axioms and definitions introduced in this chapter.

    a. Prove: $3=2+(0+1)$

    b. Prove: $(2+0)+0=1+1$

    c. Prove: $-2+(0+2)=0$

    d. Prove: $5=2+(1+2)$

    e. Prove: $2+(2+-2)=2$

IV. Give a formal proof of the following argument within the theory of arithmetic developed up to this point.

    Prove: $\neg(y=-4 \;\;\rightarrow\;\; x=-2)$

    (1) $\neg(y=(2+0)+1 \;\;\lor\;\; y+-x\neq -5)$

    (2) $x=-(-1+3) \;\;\rightarrow\;\; y=-4+(3+4)$

    (3) $y+-x=3+(-5+-3) \;\;\rightarrow\;\; y=-5+1$

CHAPTER EIGHT
UNIVERSAL GENERALIZATION

▶ 8.1 *Theorems with Variables*

You may have noticed that the proof of Theorem 36 in the exercises of the last section (7.4) seemed almost identical to the proof of Theorem 31. In fact, the only difference was that '1' replaced '2' and '2' replaced '1'. Rather than prove theorems separately for each number we may as well write them in terms of arbitrary numbers x, y, and z. In this case we specify in terms of the variables 'x', 'y', and 'z' instead of the number symbols '1', '2', and so forth. Otherwise the structure of the proofs remains unchanged.

For example, if we formulate Theorem 31 of the previous chapter in terms of 'x' and 'y', it reads

$$x + (y + (-x)) = y$$

and the proof follows precisely the second one given for Theorem 31, with 'x' replacing '1' and 'y' replacing '2'.

(1)	$x + (y + -x) = (x + y) + -x$	x/x, y/y, $-x/z$, Assoc. Axiom
(2)	$= (y + x) + -x$	x/x, y/y, Comm. Axiom
(3)	$= y + (x + -x)$	y/x, x/y, $-x/z$, Assoc. Axiom
(4)	$= y + 0$	x/x, Neg. Axiom
(5)	$= y$	y/x, Zero Axiom

But just as they stand, such theorems are not particularly helpful to us. We also want to use theorems expressed in terms of variables in the proofs of other general theorems, and for that purpose we need to be able to specify new variables for old. For instance, in the new form of Theorem 31 just given we might need to use the special case for which $y = x$, and thus to specify x/x and also x/y to obtain

$$x + (x + (-x)) = x.$$

263

In order to do this we would need to have the theorem expressed in terms of universal quantifiers to permit the use of specification:

$$(\forall x)(\forall y)(x + (y + -x = y)).$$

From this, Theorems 31 and 36 can be obtained by universal specifications

(THEOREM 31)· $1 + (2 + -1) = 2$　　　　$1/x,\ 2/y$
(THEOREM 32)　$2 + (1 + -2) = 1$　　　　$2/x,\ 1/y$

Furthermore, an unlimited number of additional theorems can be obtained by different specifications.

The rule for adding the universal quantifiers is called Universal Generalization and we abbreviate it as 'UG'. Its logical justification is straightforward: whatever we can assert, or establish from premises, for any *arbitrary* object must hold for *every* object.

The rule of *Universal Generalization* is

From formula S infer $(\forall v)(S)$.

There are some conditions under which this rule cannot be applied. (See footnote on page 267.) However, in the problems in this book these conditions do not arise at any point wherein we would need to apply UG.

We would apply the rule to the proof given on page 263 by adding as line (6).

(6)　$(\forall x)(\forall y)(x + (y + -x) = y)$　　　　UG 5

Universal generalization may also be used to solve the inference problem mentioned at the beginning of Chapter 5. There we stated that the following argument seemed intuitively sound but we could not derive the conclusion with the rules yet available to us.

All birds are animals.
All robins are birds.
Therefore, all robins are animals.

Symbolizing the premises as before, the argument is a simple one.

(1)　$(\forall x)(Bx \rightarrow Ax)$　　　　P
(2)　$(\forall x)(Rx \rightarrow Bx)$　　　　P

(3) $Rx \rightarrow Bx$	x/x 2
(4) $Bx \rightarrow Ax$	x/x 1
(5) $Rx \rightarrow Ax$	HS 3, 4
(6) $(\forall x)(Rx \rightarrow Ax)$	UG 5

Line (6) is, of course, the symbolic translation of the conclusion 'All robins are animals'.

As a second example, consider the following argument.

> No fish are mammals.
>
> All dogs are mammals.
>
> Therefore, no fish are dogs.

Define: $Fx \leftrightarrow x$ is a fish,

$Mx \leftrightarrow x$ is a mammal,

and $Dx \leftrightarrow x$ is a dog.

Prove: $(\forall x)(Fx \rightarrow \neg Dx)$

(1) $(\forall x)(Fx \rightarrow \neg Mx)$	P
(2) $(\forall y)(Dy \rightarrow My)$	P
(3) $Fx \rightarrow \neg Mx$	x/x 1
(4) $Dx \rightarrow Mx$	x/y 2
(5) $\qquad\quad Fx$	P
(6) $\qquad\quad \neg Mx$	PP 3, 5
(7) $\qquad\quad \neg Dx$	TT 4, 6
(8) $Fx \rightarrow \neg Dx$	CP 5, 7
(9) $(\forall x)(Fx \rightarrow \neg Dx)$	UG 8

Notice that we wrote the second premise in line (2) in terms of the variable 'y'. In the example about robins the second premise was symbolized using 'x'. We have shown both alternatives to indicate that *either* is logically correct.

EXERCISE 1

A. Give derivations to show the following arguments are valid.

1. All snakes are reptiles.
   All reptiles are vertebrates.
   Therefore, all snakes are vertebrates.

2. No violins are woodwinds.
   All oboes are woodwinds.
   Therefore, no violins are oboes.
3. All royalists are monarchists.
   No democrat is a monarchist.
   Therefore, no democrat is a royalist.
4. All logicians are clever persons.
   No clever persons are easily fooled.
   Therefore, no logician is easily fooled.
5. All station wagons are automobiles.
   All automobiles are vehicles.
   Therefore, all station wagons are vehicles.
6. No mammals are birds.
   All snipes are birds.
   Therefore, no snipes are mammals.
7. No cats are canines.
   All dogs are canines.
   Therefore, no cats are dogs.
8. All roses are plants.
   All plants are living things.
   Therefore, all roses are living things.
9. All drums are percussion instruments.
   All tympani are drums.
   Therefore, all tympani are percussion instruments.
10. All sonnets are poetry.
    No legal document is poetry.
    Therefore, no legal document is a sonnet.
11. $(\forall x)(\forall y)(\forall z)(x > y \quad \& \quad y > z \quad \rightarrow \quad x > z)$
    $(\forall x)(x + 1 > x)$
    Therefore, $(\forall x)(x + 3 > x)$

B. Rewrite in terms of variables the Theorems 5, 10, 15, and 31 through 42 of Chapter 7, by replacing different numerals by different variables. Now add the appropriate universal quantifiers at the beginning. Which of the resulting sentences are true in arithmetic? For those that are true give proofs.

C. Give complete formal proofs of problems 1, 5, 7, 9, 11, 14, 15, and 20 of Exercise 1 of Chapter 5.

▶ 8.2 *Theorems with Universal Quantifiers*

With universal generalization now available we are in a position to prove some fundamental theorems of a general nature. For ease of reference we restate here the four axioms introduced in Chapter 7.

Commutative Axiom:	$(\forall x)(\forall y)(x+y=y+x)$
Associative Axiom:	$(\forall x)(\forall y)(\forall z)((x+y)+z=x+(y+z))$
Axiom for Zero:	$(\forall x)(x+0=x)$
Axiom for Negative Numbers:	$(\forall x)(x+(-x)=0)$

Because the theorems we prove in this section do not depend on those proved in Chapter 7, we shall renumber starting again with 1. The first theorem asserts what is known as the left-hand *cancellation* law for addition. Explanation of the theorem follows its proof.

THEOREM 1. $(\forall x)(\forall y)(\forall z)(x+y=x+z \;\;\to\;\; y=z)$

Proof.

(1)	$x+y=x+z$	P*
(2)	$y=y+0$	y/x, Zero Axiom
(3)	$=0+y$	$y/,\ 0/y$, Comm. Axiom
(4)	$=(x+(-x))+y$	x/x, Neg. Axiom
(5)	$=(-x+x)+y$	$x/x,\ -x/y$, Comm. Axiom
(6)	$=-x+(x+y)$	$-x/x,\ x/y,\ y/z$, Assoc. Axiom
(7)	$=-x+(x+z)$	Line (1)
(8)	$=(-x+x)+z$	$-x/x,\ x/y,\ z/z$, Assoc. Axiom
(9)	$=(x+-x)+z$	$-x/x,\ x/y$, Comm. Axiom

---

* The conditions under which universal generalization cannot be applied arise where, as here, an additional premise is introduced containing variables without quantifiers. UG cannot be applied in this subordinate proof. However, there is no need to apply it in the subordinate proof. Line (13) is no longer in the subordinate and so does not depend on the added premise. UG may be applied. Fortunately, premises containing unquantified variables do not normally occur in arguments where we would want to apply UG. This includes subordinate proofs where such premises do often occur but where normally there is no need to apply UG.

(10)	$=0+z$	x/x, Neg. Axiom
.1)	$=z+0$	$0/x$, z/y, Comm. Axiom
(12)	$=z$	z/x, Zero Axiom
(13)	$x+y=x+z \;\rightarrow\; y=z$	CP 1, 13
(14)	$(\forall x)(\forall y)(\forall z)(x+y=x+z \;\rightarrow\; y=z)$	UG 13

To understand just what this theorem implies, let us first look at an instance of it.

$$3+y=3+4 \;\rightarrow\; y=4 \qquad 3/x, y/y, 4/z, \text{Th. 1}$$

The argument corresponding to this conditional is

Premise: $3+y=3+4$
Conclusion: $y=4$

The conclusion results from the premise if the two 3's to the left of the plus signs are cancelled out. To obtain the conditional '$x+y=x+z \;\rightarrow\; y=z$', a conditional proof is used. This requires the added premise '$x+y=x+z$'. Note that the additional premise of line (1) could have been introduced later, but it was more convenient to put it at the beginning and not break into the string of identities. Note also that line (1) is used only at one point—line (7). To prove a cancellation law we need somehow to reduce the length of expressions—to go from '$x+y=x+z$' to '$y=z$'. Two axioms make such a reduction, the Axiom for Zero and the Axiom for Negative Numbers. The strategy of proof is to apply the Axiom for Negative Numbers first to replace 'x' by '0' and then to remove '0' from '$0+z$'. Of course, the other two axioms are used repeatedly in carrying out this strategy.

The cancellation law may be used to prove the familiar fact that the negative of the negative of a number is the number itself.

THEOREM 2. $(\forall x)(-(-x)=x)$

Proof.

(1)	$x+-x=0$	x/x, Neg. Axiom
(2)	$-x+x=0$	x/x, $-x/y$, Comm. Axiom 1
(3)	$-x+-(-x)=0$	$-x/x$, Neg. Axiom
(4)	$-x+-(-x)=-x+x$	I 1, 3
(5)	$-x+-(-x)=-x+x \;\rightarrow\; -(-x)=x$	$-x/x$, $-(-x)/y$, x/z, Th. 1

(6) $-(-x)=x$ PP 4, 5
(7) $(\forall x)(-(-x)=x)$ UG 6

The proof of Theorem 3 follows. Note that its main sentential connective is an *equivalence*. In this situation it is necessary to break the proof into two parts, proving first one implication and then the other, each by conditional proof.

THEOREM 3. $(\forall x)(\forall y)(-x=y \ \leftrightarrow \ x+y=0)$.

Proof.
(1) $-x=y$ P
(2) $x+-x=0$ x/x, Neg. Axiom
(3) $x+y=0$ I 1, 2
(4) $-x=y \ \rightarrow \ x+y=0$ CP 1, 3
(5) $x+y=0$ P
(6) $x+-x=0$ x/x, Neg. Axiom
(7) $x+y=x+-x$ I 5, 6
(8) $x+y=x+-x \ \rightarrow \ y=-x$ $x/x, y/y, -x/z$, Th. 1
(9) $-x=y$ PP 7, 8
(10) $x+y=0 \ \rightarrow \ -x=y$ CP 5, 9
(11) $-x=y \ \leftrightarrow \ x+y=0$ LB 4, 10
(12) $(\forall x)(\forall y)(-x=y \ \leftrightarrow \ x+y=0)$ UG 11

Theorem 3 says that $-x$ is the only number that can be added to x to obtain the sum 0.

At this point we define the binary operation of subtraction in terms of addition and the negative operation. As was pointed out in Chapter 7, from a logical standpoint such a definition acts as an additional premise. It has the same status as an axiom; it may be assumed and used in all proofs as a premise.

DEFINITION 1. $(\forall x)(\forall y)(x-y=x+(-y))$

We first prove that subtracting the negative of a number is the same as adding the number itself. The proof depends on essential use of Theorem 2, as well as Definition 1.

THEOREM 4. $(\forall x)(\forall y)(x-(-y)=x+y)$

Proof.

(1)	$x-(-y)=x+ -(-y)$	$x/x, -y/y$, Def. 1
(2)	$=x+y$	y/x, Th. 2
(3)	$(\forall x)(\forall y)(x-(-y)=x+y)$	UG 2

We next want to prove that $x-0=x$, but to do so it is convenient first to prove that the negative of 0 is 0. This proof uses essentially Theorem 3.

THEOREM 5. $-0=0$

Proof.

(1)	$-0=0 \quad \leftrightarrow \quad 0+0=0$	$0/x, 0/y$, Th. 3
(2)	$0+0=0$	$0/x$, Zero Axiom
(3)	$0+0=0 \quad \rightarrow \quad -0=0$	LB 1
(4)	$-0=0$	PP 2, 3

The proof that $x-0=x$ we leave as an exercise, but it is easy with Theorem 5 available. Additional theorems are given in the exercises.

EXERCISE 2

A.   Prove the following theorems.

THEOREM 6. $(\forall x)(x-0=x)$
THEOREM 7. $(\forall x)(0-x= -x)$
THEOREM 8. $(\forall x)(\forall y)(\forall z)((x-y)+(y-z)=x-z)$
THEOREM 9. $(\forall x)(\forall y)(-x=y \quad \leftrightarrow \quad x=-y)$
THEOREM 10. $(\forall x)(\forall y)(\forall z)(x+y=z+y \quad \rightarrow \quad x=z)$
THEOREM 11. $(\forall x)(\forall y)((-y+ -x)+(x+y)=0$
THEOREM 12. $(\forall x)(\forall y)(-(x+y)= -x+ -y)$
THEOREM 13. $(\forall x)(\forall y)(\forall z)(\forall w)((x-y)+(z-w)=(x+z)-(y+w))$
THEOREM 14. $(\forall x)(\forall y)(\forall z)(\forall w)((x-y)-(z-w)=(x+w)-(y+z))$
THEOREM 15. $(\forall x)(\forall y)(\forall z)(x-y=z \quad \leftrightarrow \quad x-z=y)$
THEOREM 16. $(\forall x)(\forall y)(\forall z)(x+y=z \quad \leftrightarrow \quad x-z= -y)$
THEOREM 17. $(\forall x)(\forall y)(\forall z)(\forall w)(x-y=z-w \quad \leftrightarrow \quad x+w=y+z)$

B.   Give formal proofs of the following arguments within the theory of arithmetic of this chapter and the definitions of the integers if needed.

1. Prove: $-3 < -2$
   (1) $(\forall x)(x < x + 1)$
2. Prove: $x + x = 0$
   (1) $x = 0$
3. Prove: $x = -4$
   (1) $x + 5 = 1$
4. Prove: $(\forall x)(x < 0 \ \leftrightarrow \ 0 < -x)$
   (1) $(\forall x)(\forall y)(\forall z)(y < z \ \rightarrow \ (x + y < x + z))$
5. Prove: $(\forall x)\neg(x < x)$
   (1) $(\forall x)(\forall y)(x < y \ \rightarrow \ \neg(y < x))$

INDEX

B C D E F G H I J 0 6 9 8 7 6
PRINTED IN THE UNITED STATES OF AMERICA